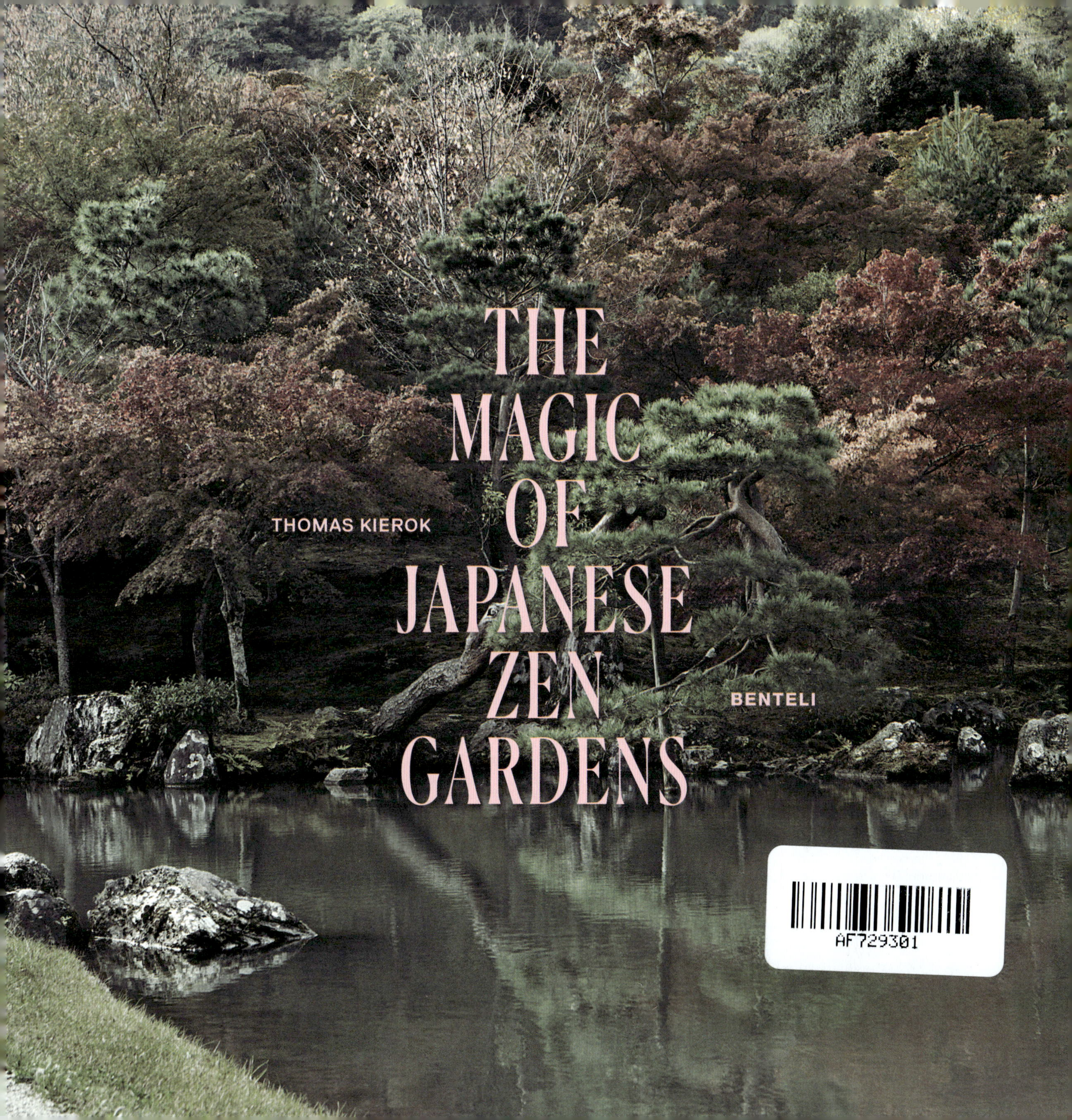

THE MAGIC OF JAPANESE ZEN GARDENS

THOMAS KIEROK

BENTELI

To live and to be aware of transience in every moment is very important. Transience – *mujō* – is the essence of the world. It is very important for us humans to feel *mujō* to be able to recognize the richness of life. The seasonal changes allow us to feel this immediately.

Masuno Shunmyō

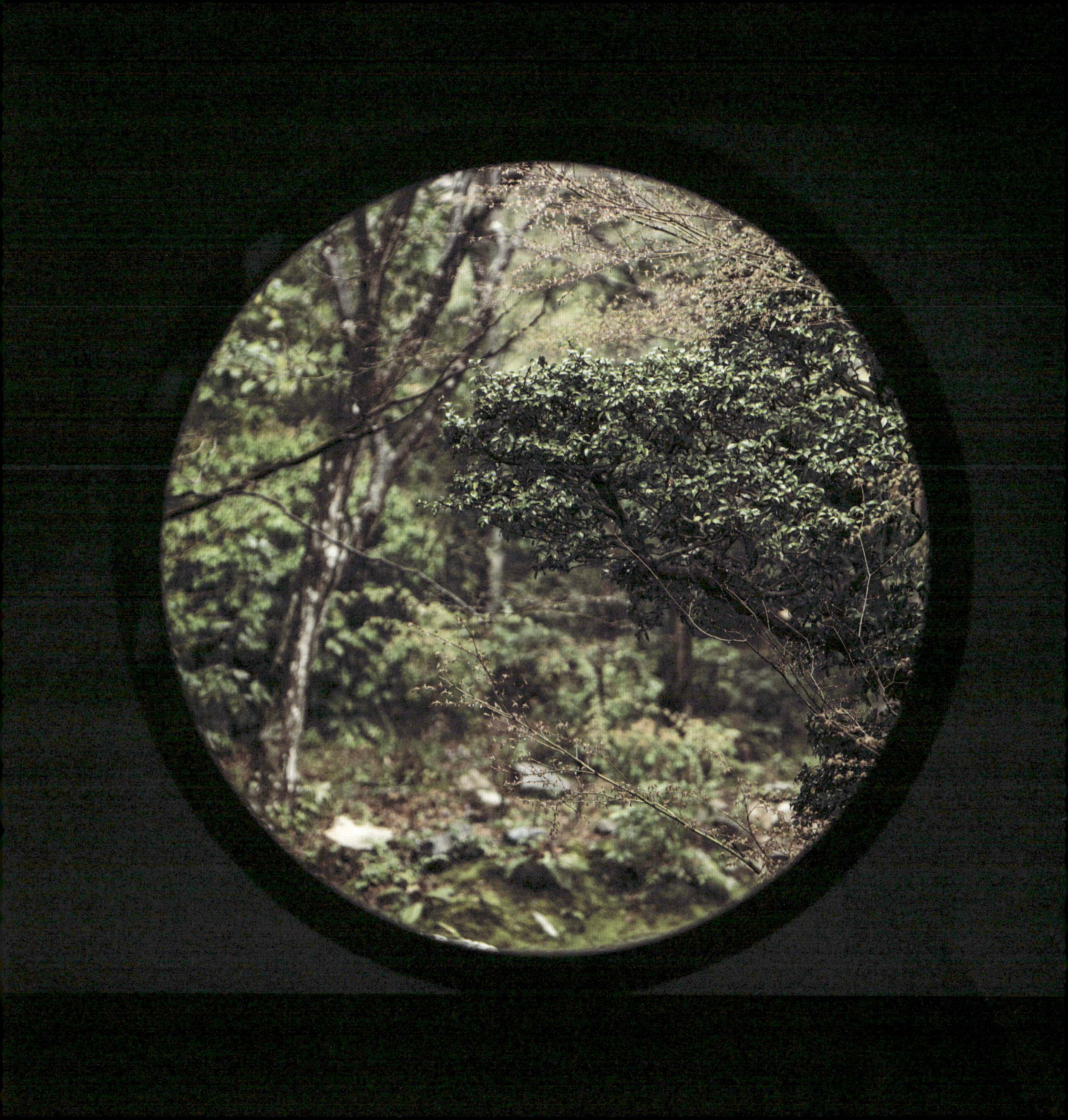

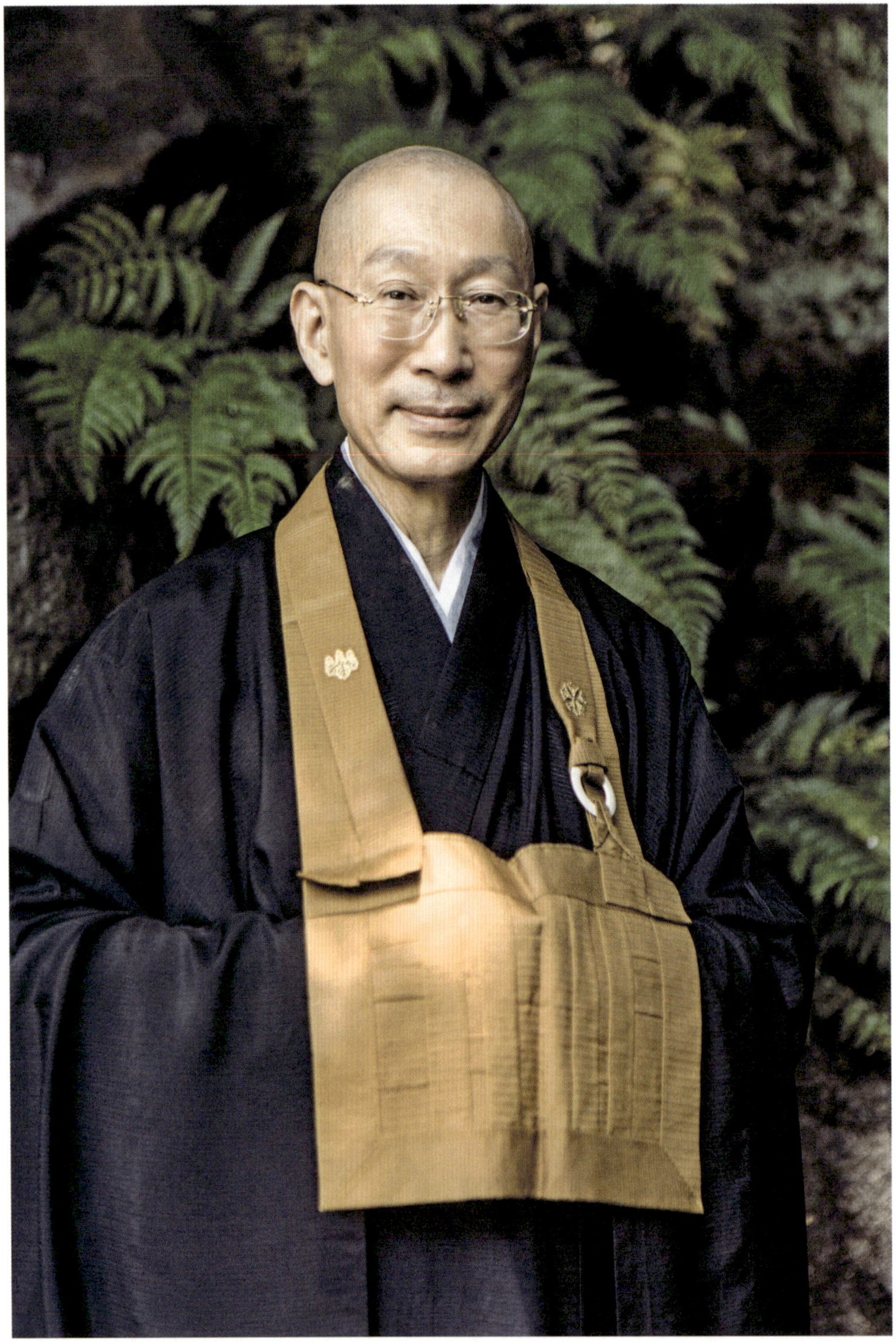

PREFACE

It is so wonderful that Mr. Thomas Kierok's photography book *The Magic of Japanese Zen Gardens* is published. The book showcases our world uniquely captured through the eye of a photographer from abroad. I believe that this book will allow an international audience to become interested in Zen gardens. The photographs were taken in different seasons, with each photograph reflecting the drastic changes of the garden throughout the year. The images are all very refreshing, even surprising. These days, people call any garden a Zen garden, as long as it has stones and white pebble, regardless of its other content. This is an ongoing trend not only in Japan, but even more so outside Japan. The concept of a Zen garden is not that simple.

So what is a Zen garden? Would every garden in a Zen temple be considered a Zen garden? – NO. Because it cannot be a Zen garden if it does not express the designer's Zen state of mind put into three dimensional art. Even if the garden is designed by a well-known designer with a highly sophisticated technique, it cannot be called a Zen garden if it was created based on a fixed view. It has to reflect the designer's own essence. All Zen art is an expression of the artists' own experience of meeting their own nature in visual form. Therefore, as long as the piece of art is created in that way, it can be true Zen art, even if that person is not an ordained priest. In Zen art, the method and technique is not important, while the creator's spirituality is. Only a garden that expresses the state of understanding attained by practice can be called a Zen garden.

Every garden featured in this book is a true Zen garden in Kyoto. A part of each one is captured throughout the seasons in pictures that show stone gardens and abundant nature. In Zen, practitioners try to find infinite teaching and truth in nature. Nature is considered to contain Buddha himself and *Dharma* teachings. There is an emphasis in Zen practice to meditate in nature. The most ideal life for a practitioner is to live lead by the truth that life is simply given to us – and to be aware that infinite teachings and truth are accessible all the time. Only a garden in the high state of the art, using nature as a symbol or extracting it into the abstract form can be called a Zen garden. In this sense, all the gardens introduced here are authentic Zen gardens. I would like to congratulate Mr. Thomas Kierok for publishing this book, and thank him for asking me to write this introduction.

Two palms together (In *Gasshō*)
July 14th, 2022
Masuno Shunmyō, Abbott
Tokuyuzan Zuiun-In, Kenchō-ji temple

INITIAL REMARKS

When Thomas Kierok told me for the first time a few years ago – significantly in a *Zen dojo* in the middle of Berlin – about his project to photograph the gardens of Japan, I initially asked myself the question: is there really a need for another photography book on Japanese gardens? Zen gardens in particular? That was shortly before the first of four journeys (during the four seasons) that the photographer made to Japan in the following three years, explicitly to visit these gardens, mainly in Kyoto. Since he not only impressed me by the stringency with which he approached his project, but also by the conversations that have taken place since then, and above all the photographic results of his travels, my initial question was soon resolved. But what makes these photographs so special? Thomas Kierok's approach is that of a professional portrait photographer. Prerequisite for a good portrait is the engagement with the subject and a sense for the ongoing communication, which is certainly mostly non-verbal. In other words, the dynamic process of a primarily inner approach. If this succeeds, something essential can communicate itself to the resulting picture and, paradoxically, the seemingly fixed picture can then in turn convey something of this dynamic to the viewer as well.

These garden pictures are, so to speak, portraits resulting from a fascination with the occasionally almost magical or enchanted landscape worlds and propelled by an in-depth engagement that can both make the enormous power of these gardens palpable as well as potentially constitute the onset of a subtle interchange that may reach unimaginable depths at times.

It seems to me that an essential part of this process resonates in Kierok's pictures, because they are about a journey directed inside! Similarly when a master of the art creates a Zen garden, he/she also embarks on such a journey, the beginning of which is an immersion into the given landscape, from which the compositional approaches then emerge, so to speak between gentle integration and, in part, abruptly contrasting interventions. For example, juxtaposing a lush wild mountain world background with extremely reduced gravel surfaces, which nevertheless appear as a unit related to each other. When it comes to the rock settings, it is important just to listen: the rocks will tell you where they belong ...

Bernd A. Schellhorn

ZEN GARDENS STAND FOR HARMONY, SLOWNESS AND SIMPLICITY

My journey to the traditional Japanese Zen gardens in Kyoto is a personal one. Like many travelers, I arrived at one of these world-famous gardens with no intentions in mind. Touched by the beauty and grace, they awakened in me a great sense of friendship and serenity. This present feeling of happiness in the Zen gardens is a connection to my childhood spent in the nature of Hamburg Gartenstadt.

Japanese culture, martial arts, Japanese characters and the Zen philosophy have interested and inspired me since my early youth.

The 7 different characteristics of Zen art and Zen gardens are:

ASYMMETRY (*FUKINSEI*)

SIMPLICITY (*KANSO*)

MATURITY, AUSTERITY (*KOKO*)

NATURALNESS (*JINNEN*)

SUBTLY PROFOUND (*YŪGEN*)

FREE FROM CONVENTIONS (*DATSUZOKU*)

TRANQUILITY (*SEIJAKU*)

Kokoro is the heart, spirit and soul of the elements of the garden that connects with the designer and the viewer. Zen gardens delight the sight and are a mirror of the heart and the soul. The viewers are invited to discover the magic of the gardens and their own Kokoro.

Thomas Kierok

Zen gardens are places where people can intensely perceive themselves. There is the beauty of the garden, but people should not see the garden that stretches out before them as something objective, but as part of themselves. I want them to see the garden they are looking at and their own selves as one.

Things are as they appear. The stones, the trees, the water in a garden – everything is there as itself. In Zen, this is called *hotoke no*, which literally means *the Buddha himself*. This way of seeing is the same for Japanese and non-Japanese.

Masuno Shunmyō

INTRODUCTION

Zen gardens have fascinated me for many years. Then I discovered my first Zen garden almost on my doorstep in the Berlin "Gardens of the World". In 2014 I went to Japan for the first time to meditate in a Zen monastery. By chance – I had received a cooking course as a gift – I met a Japanese cook. She told me about Saigaku, a Japanese monk who lives in Berlin. He traveled to Tokyo a short time later accompanied by a small group and from there to a small monastery near Eihei-ji, the famous temple established by the founder of the *Sōtō* Zen tradition, Dōgen Zenji. I joined this group and spent five days at Eihei-ji. Since it was close to Kyoto, I naturally went to see the famous Zen gardens in Kyoto. Without knowing much about them, these gardens touched me very strongly. Shortly before my return flight, I discovered the book *Zen Gardens: The Complete Works of Shunmyō Masuno, Japan's Leading Garden Designer* in a bookstore at the Tokyo airport. In it I learned that this Masuno Shunmyō had also designed the Berlin Zen Garden. So I already knew him without knowing him. At that moment I knew that I would return to Japan and continue photographing Zen gardens. On my second trip to Japan, I was able to meet Masuno Shunmyō in person. He is an 18th generation Zen priest who lives and works in his birthplace of Yokohama at Kenkohji Temple, which with its old wooden buildings seems like a timeless oasis of peace and contemplation. He works as an architect and Zen garden designer on the one hand and as a priest on the other. He also began drawing at an early age and eventually became an apprentice to one of the most famous Japanese Zen landscape architects.

At the same time, he followed in his father's footsteps and became a priest. Since then, he functions as a priest mainly on weekends and works as an architect on weekdays. Even though his work as an architect takes up most of his time, in his heart he is primarily a priest. He lives in the temple, with his office located in its reception building. He has about ten employees, with whom he works on eight to twelve projects worldwide in parallel. The development time of the projects varies between two and ten years. He is also a professor at a Japanese university. Zen philosophy, Zen practice, and the design of the gardens are one and the same for him.

In the past, one of the duties of the priesthood was to design and maintain the gardens of the temples. Today he sees himself as the last to design gardens out of religious inspiration, in Japan and also worldwide.

His mission is to spread Zen philosophy and Japanese culture to the world through his gardens. His Zen philosophy makes his designs unique and also dictates that he makes the most important deci-

sions himself, such as the selection of rocks. This selection is made intuitively on site after building models. The setting of the rocks is something deeply sacred to him. The rocks can weigh tons and are moved by crane. Masuno Shunmyō then decides intuitively and in the now how the energy will flow and, if necessary, have the rock moved. In the process, he admits, he is as excited as a child.

During my first trip to Japan, I primarily photographed the traditional Zen gardens in and around Kyoto and in Eihei-ji. I was particularly fascinated by the peculiarities of the old, traditional Zen gardens in Kyoto. Zen gardens are a visualization of Zen philosophy. They actually do not contain nature. They are a kind of paradise garden, created by the mind and hand of man. A Zen garden is so perfectly naturally staged that one does not suspect the staging. Both creating and maintaining such a garden requires mindfulness, being in the here and now. For me as a photographer, that means slower, less, and more deliberately taking photos.

My favorite garden is the famous Ryoan-ji, about the size of a tennis court and studded with a number of rocks, you can never see all at once. I stayed nearby at a temple hotel and visited the garden regularly early in the morning. After the second time, I went there just to be there, not to take pictures. I sat down, felt and looked. A relaxed feeling of happiness spread throughout me. Quite unspectacularly and nevertheless particularly I felt an otherwise so rarely experienced composure, peace and a being-with-me.

Zen gardens require me to develop a very individual view, which is an intuitive process.

I photographed Zen gardens on four of my trips to Japan, in all four seasons. This allowed me to see them from different angles. On my second trip, the gardens were much smaller in reality than in my pictures and memory. This surprised me a lot. I had a favorite picture of the garden at Tenryu-ji, one of the most famous gardens. The picture is taken through a passageway. I walked the same pathway on my second visit as I had six months earlier, and I could have sworn that the tree on it had the circumference of one meter. But in reality its circumference is only 30 centimeters.

On the one hand, of course, one's perception changes permanently, but so does one's memory. On the other hand, I may not have looked properly the first time. The second time, I had background knowledge and looked more precisely.

I photographed the Zen gardens in all four seasons and naturally wanted to capture the special magic of each season. This was easier to plan in summer and autumn than in winter and spring, as I wanted to photograph the gardens in the snow and during the cherry blossoms, respectively. However, it generally only snows once or twice in Kyoto and in some years not at all. After an acquaintance in Kyoto informed me on Wednesday that it would probably snow in Kyoto on Saturday, I flew to Osaka two days later and then took the train to Kyoto. There was no sign of snow, but shortly before Kyoto the first snowflakes fell – and that brought tears to my eyes. Also in the following days snow fell again and again, quite an exceptional amount, as I was told. Nevertheless, at the beginning I was still gripped by a certain restlessness and went to the gardens very early in the morning, because I was afraid that it could soon be over with the snowy splendor. Apart from me, there were hardly any visitors there, and so I was able to photograph completely undisturbed and, as time went by, to immerse myself more and more deeply in what was presented to me – an infinite number of different shades of gray. It was quite different in the spring. I wanted to be in Kyoto to photograph at the time of the cherry blossom – *Hanami* – and there, too, I had to plan at quite short notice, because it cannot be predicted precisely in terms of time. During these days, of course, I was mostly surrounded by crowds of tourists who wanted to experience this burst of colors in the Zen gardens. Looking back, I can say that the Zen gardens had an almost mystical effect on me at all times of the year and touched and moved me deeply. I think I experienced something of what Masuno Shunmyō describes as, "Impermanence – *mujō* – that is the essence of the world. Feeling *mujō* is very important for us humans to be able to recognize the richness of life. The seasonal changes make us feel this directly."

Sometimes I wonder where my interest in Zen gardens comes from and what it resonates in me. I have noticed that I have felt great happiness in being alone. By traveling alone, I have strongly felt the connection with nature. I originally come from the garden city of Hamburg in Wandsbek and recently found a photo showing our family's small allotment garden. I can hardly remember, but I can imagine that this meant that a garden was associated with feelings of happiness for me from an early age. Actually, I'm the total urban neurotic, having lived in Hamburg and Berlin and spent most of my time in big cities while traveling through Asia. I was surprised myself what these gardens trigger in me. Certainly, photographing Zen gardens is also work and requires a certain result-oriented concentration. At the same time, it allows me to live out a completely different side of me and, for example, to be out and about all by myself. Photographing the gardens affects me somehow – one Sunday my photo of the day was simply a stone.

Thomas Kierok

FUKINSEI
ASYMMETRY

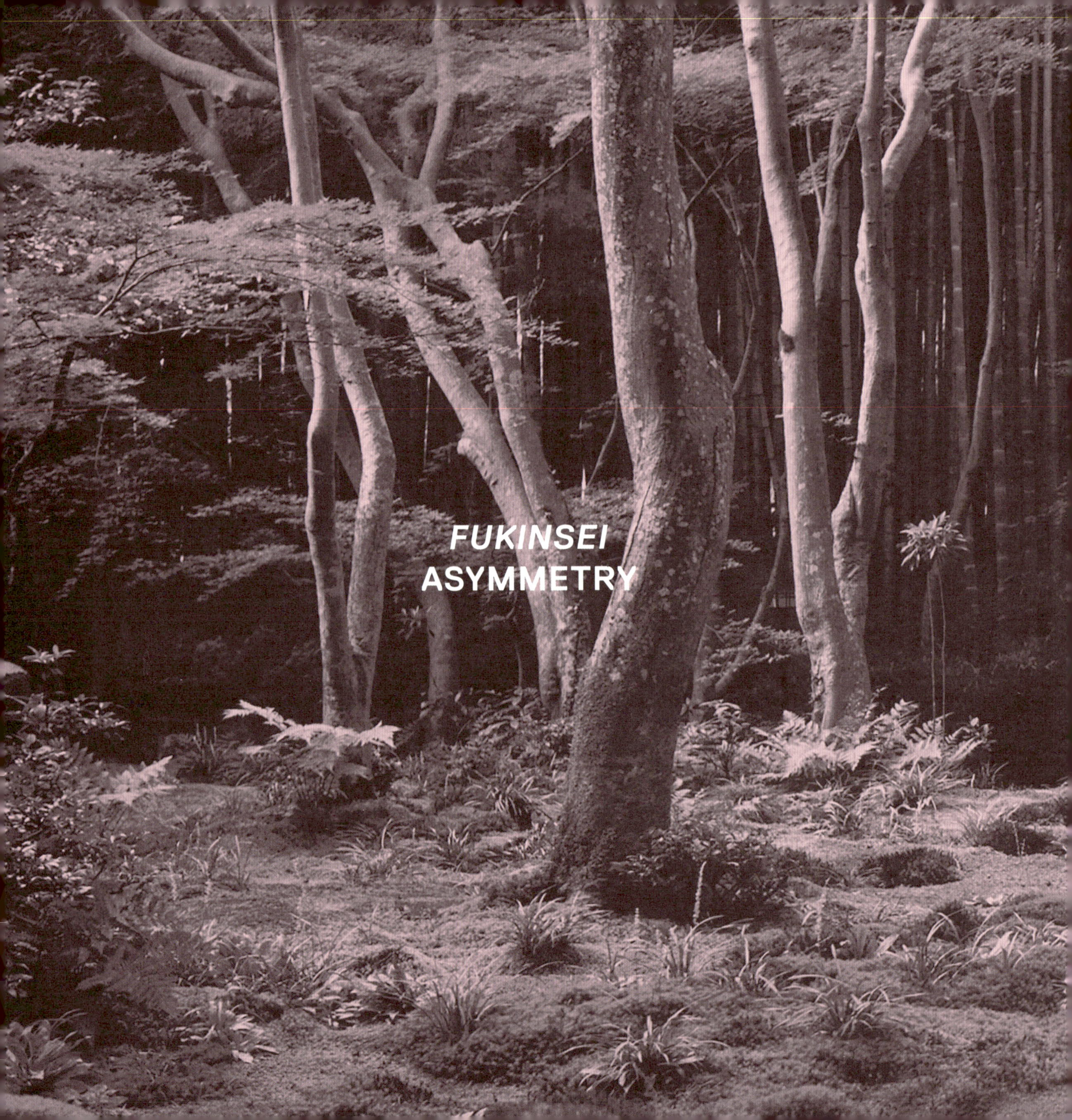

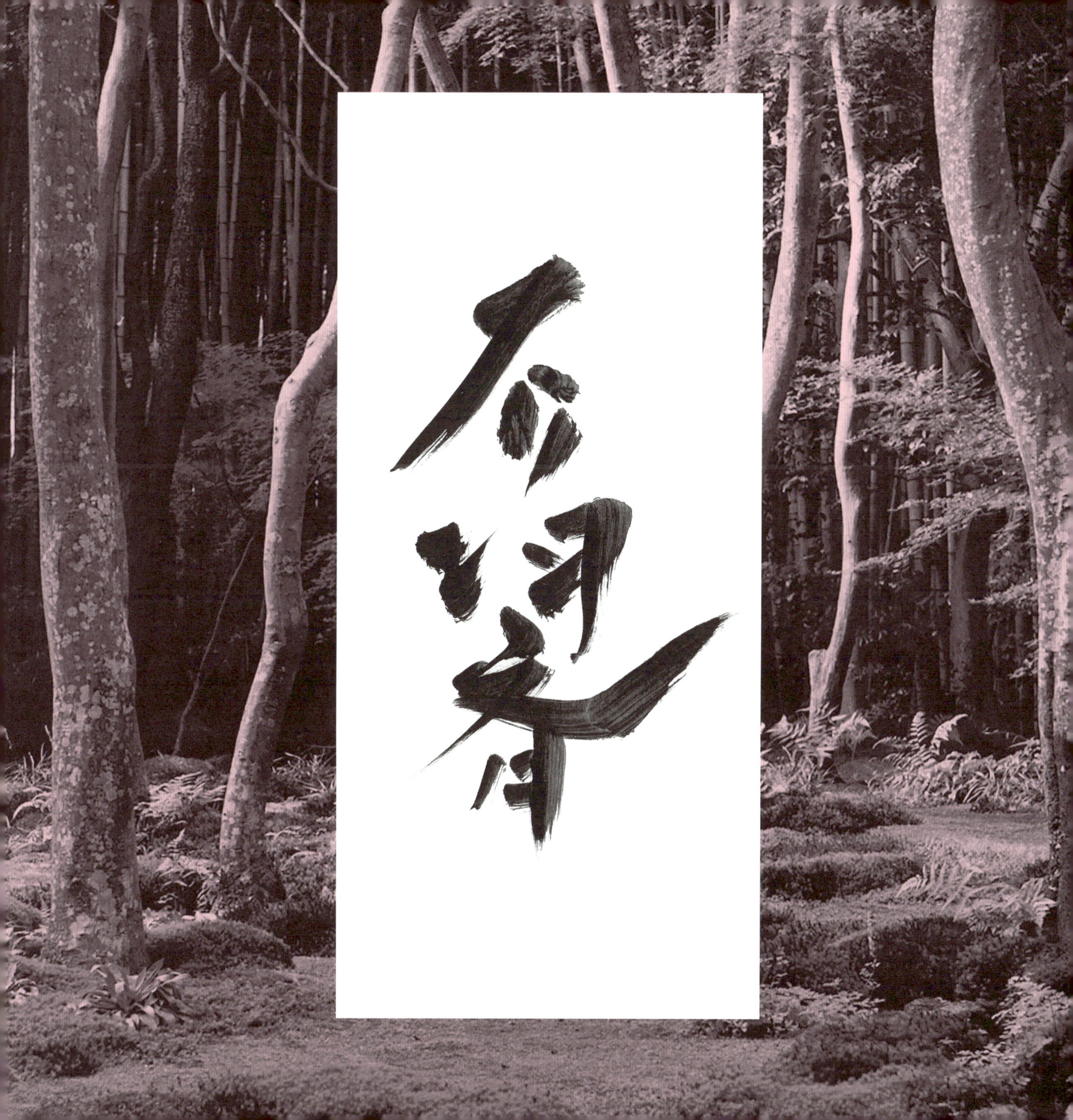

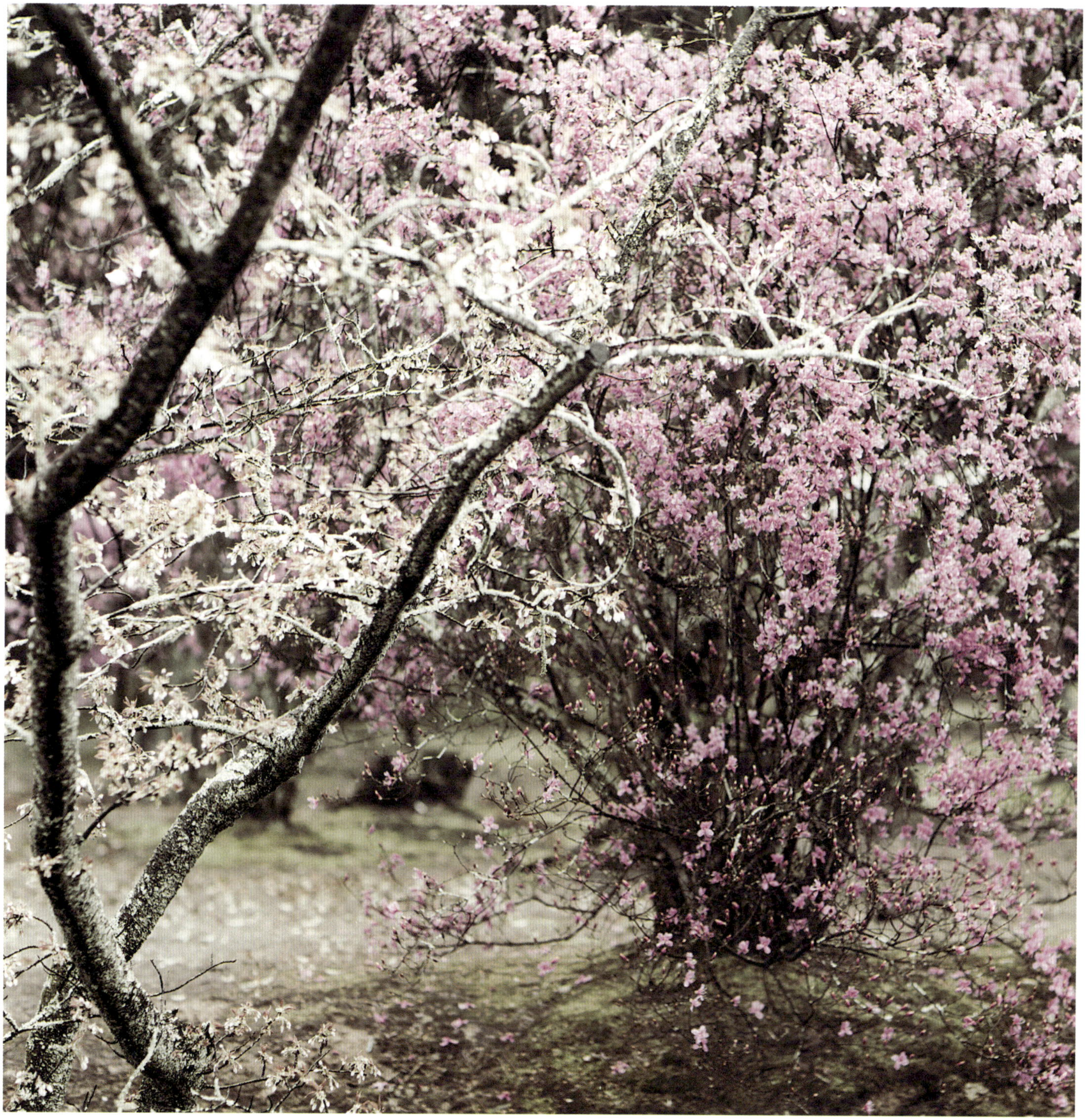

TO STUDY ZEN IS TO STUDY THE SELF.
TO STUDY THE SELF IS TO FORGET THE SELF.
TO FORGET THE SELF IS
TO BECOME ONE WITH THE UNIVERSE.

Dōgen Zenji

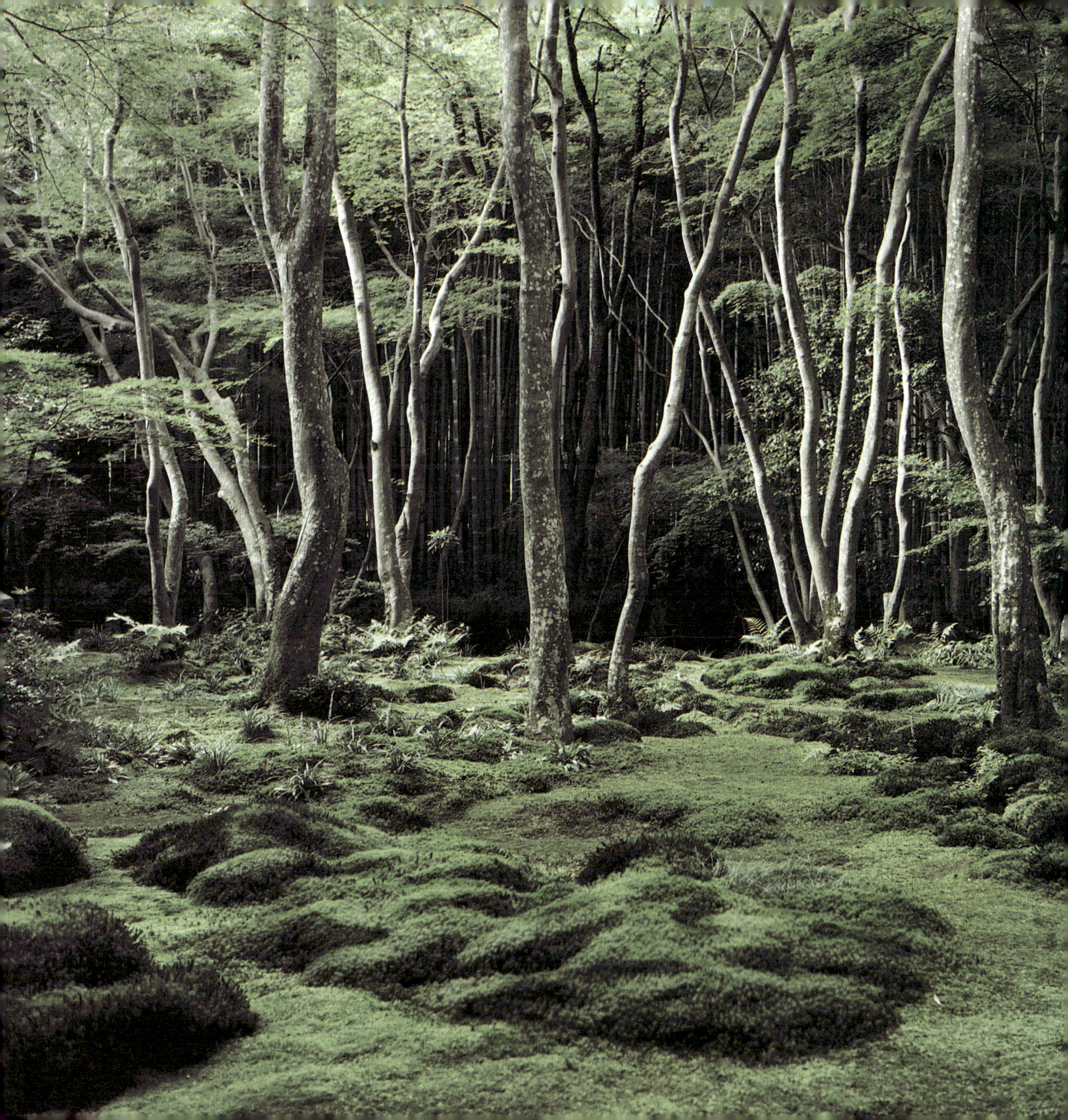

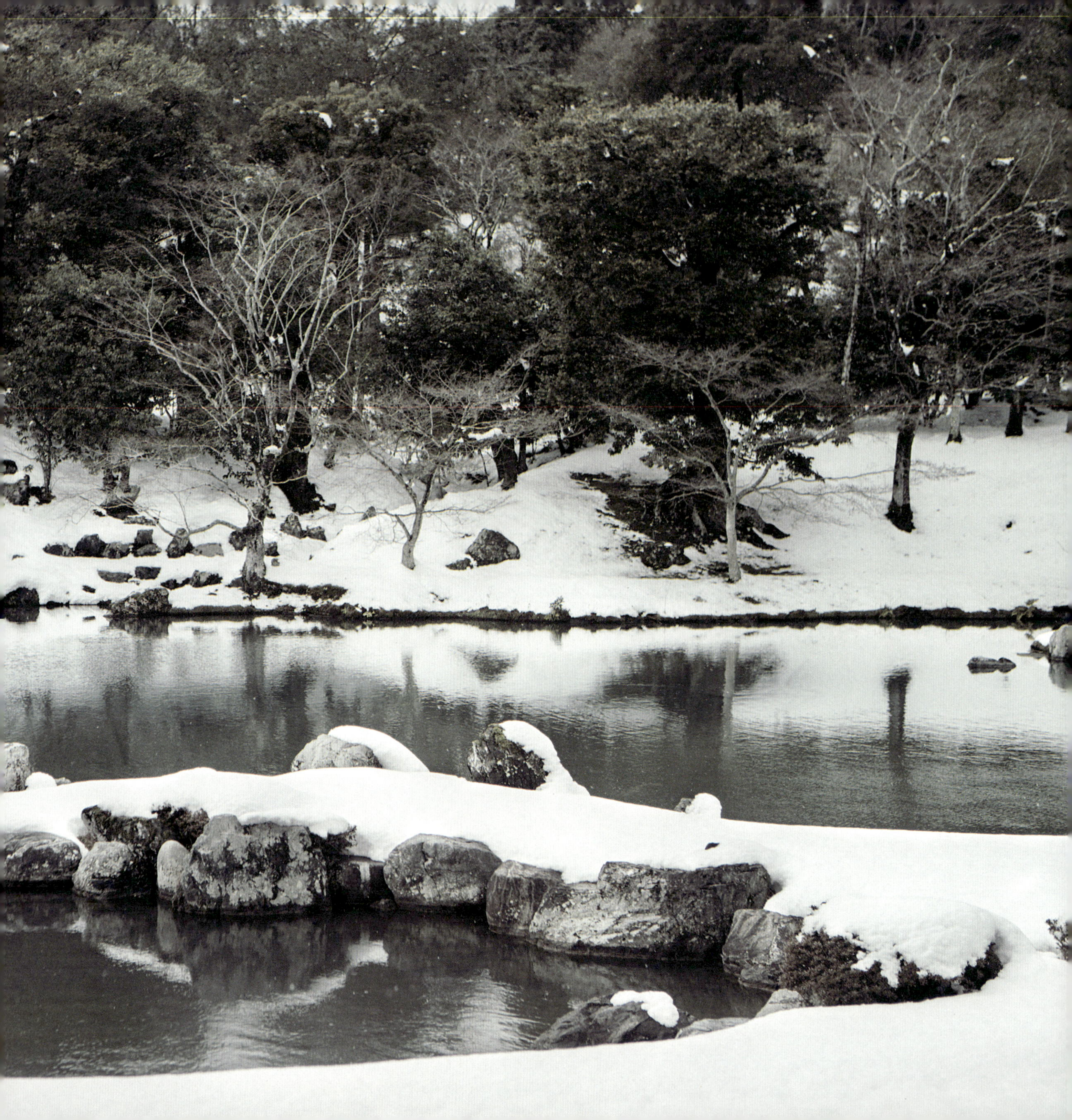

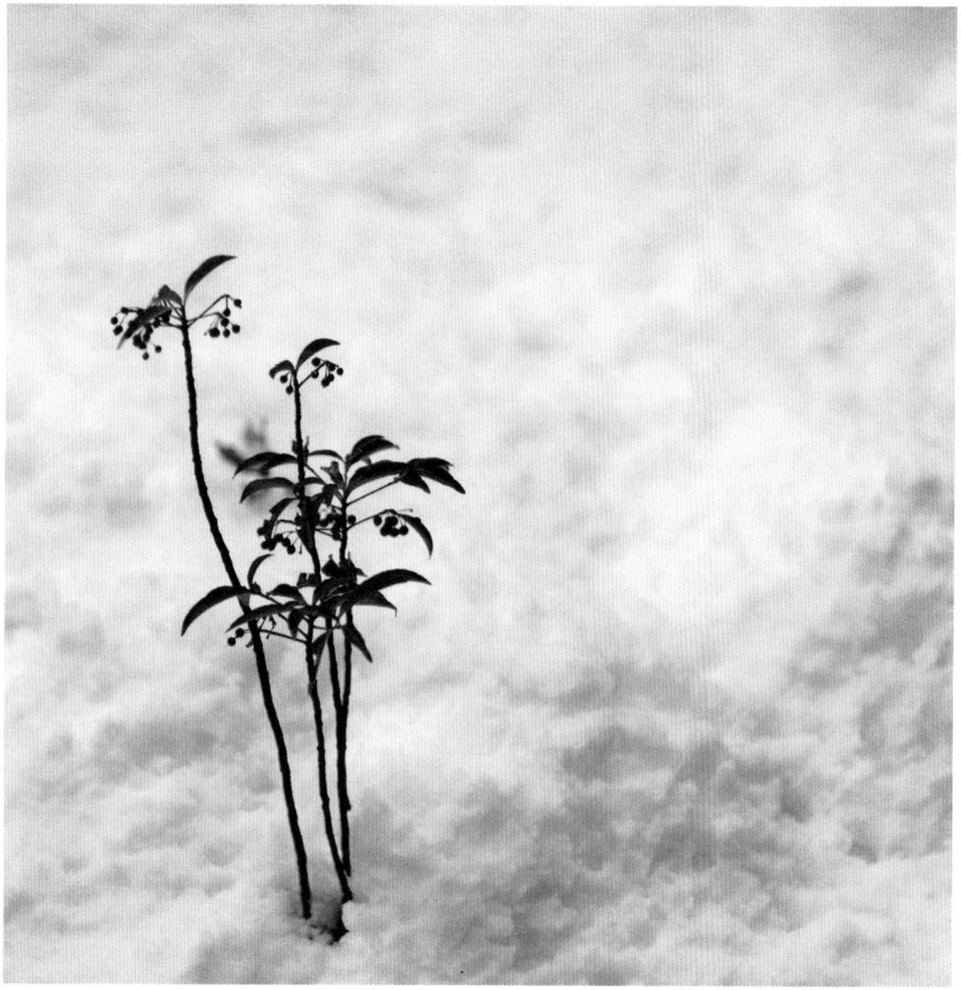

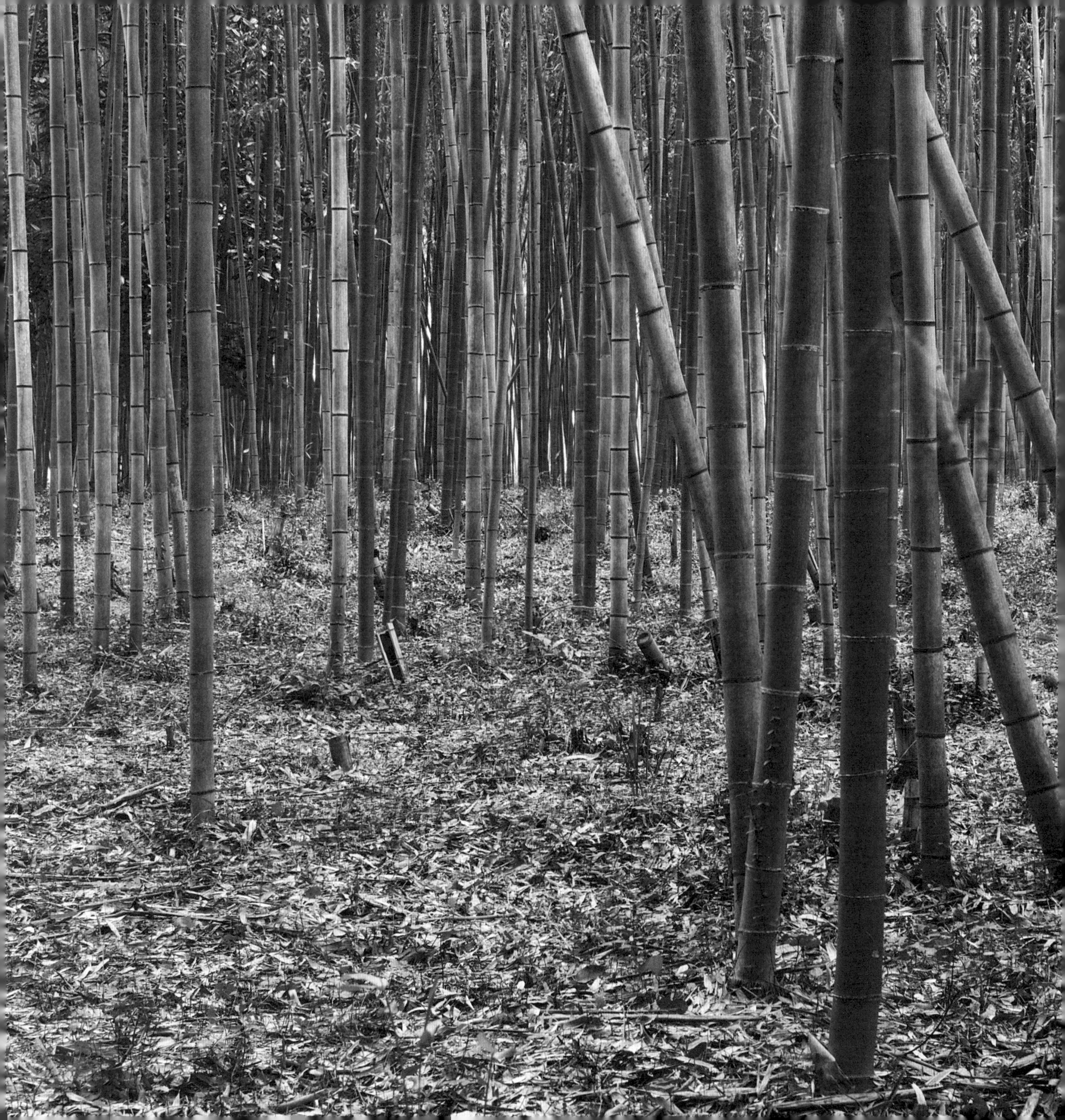

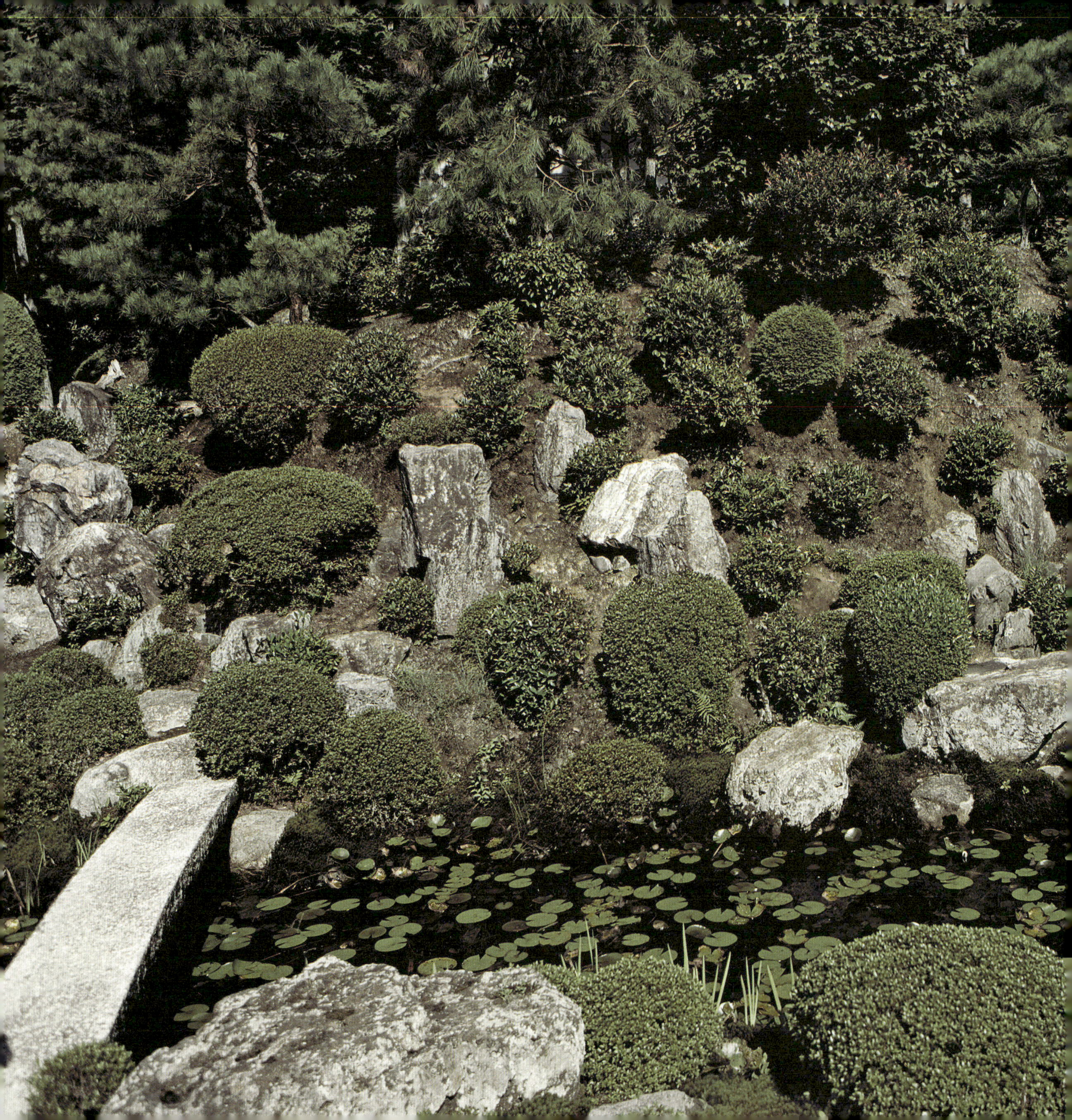

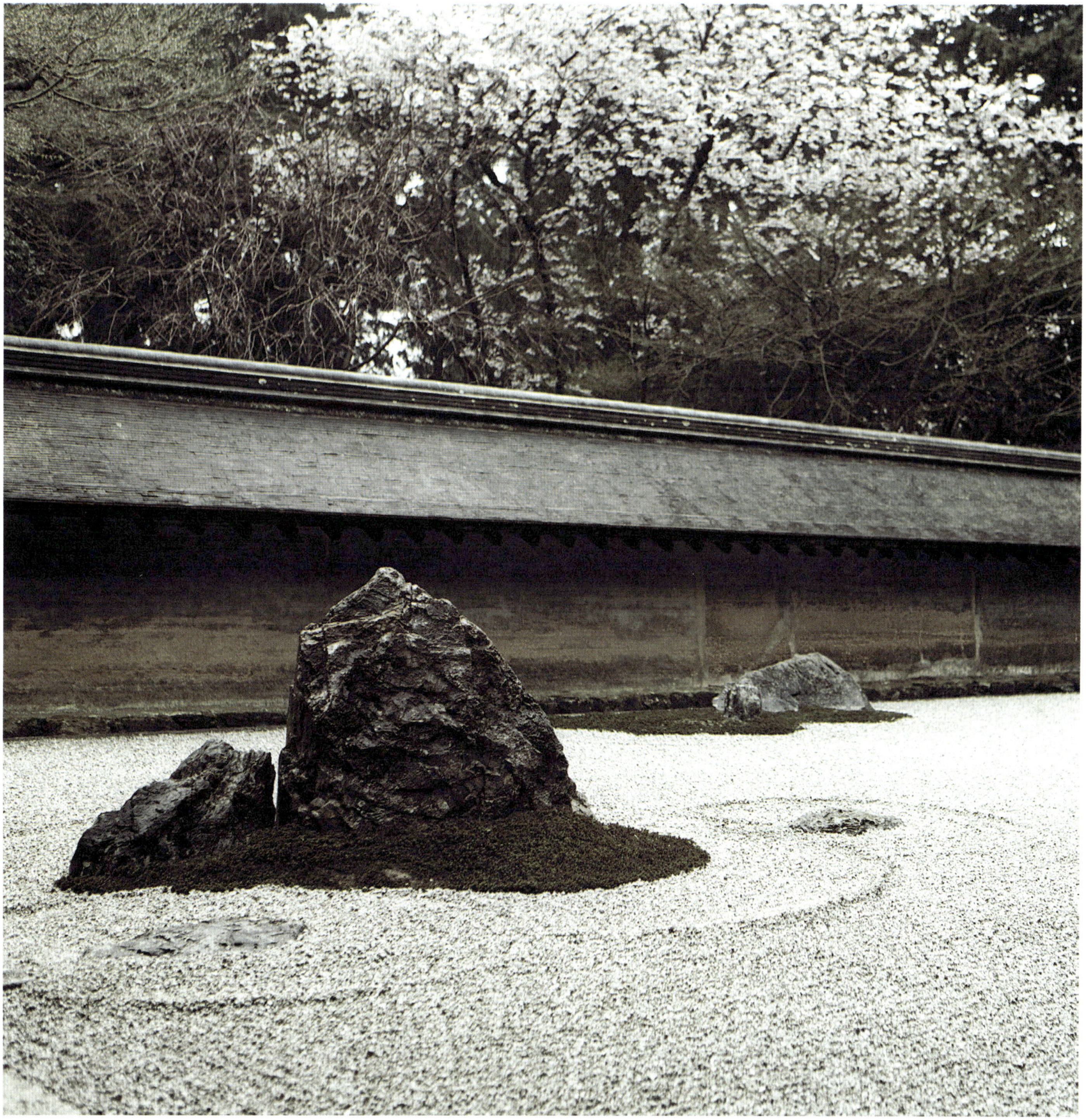

WATCHING THE MOON AT DAWN, SOLITARY, MID-SKY,
I KNEW MYSELF COMPLETELY, NO PART LEFT OUT.

Shikibu Izumi

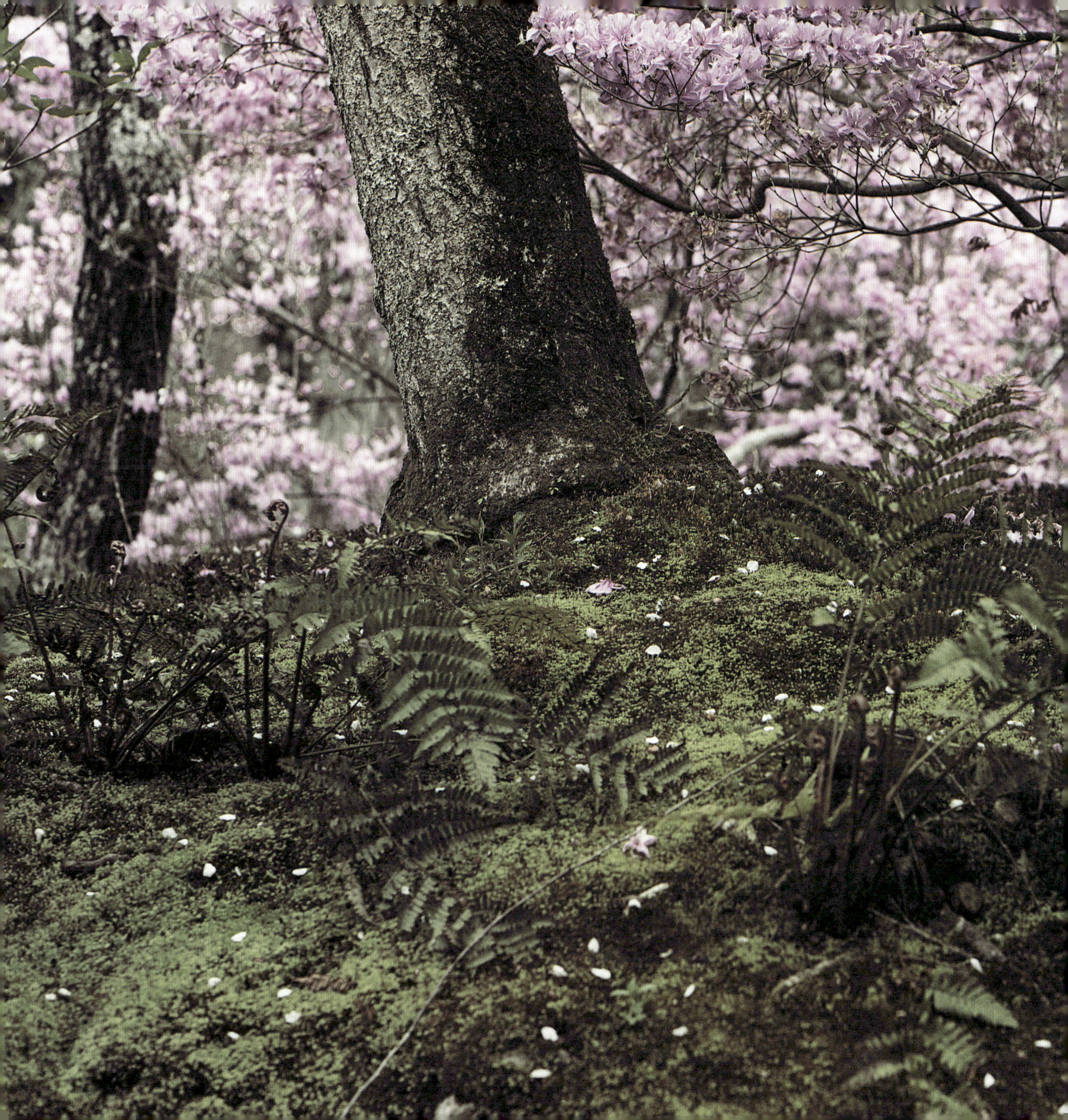

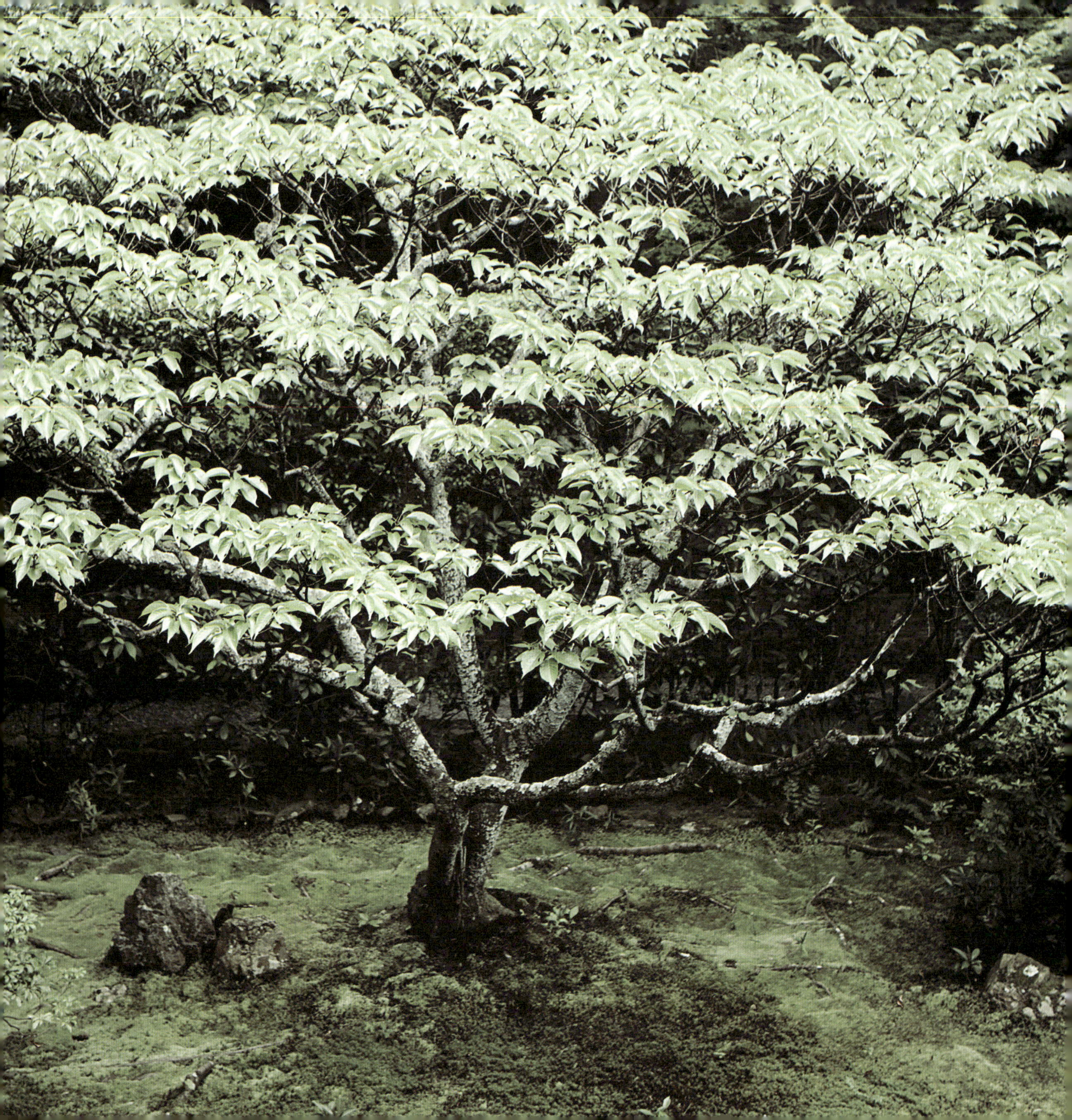

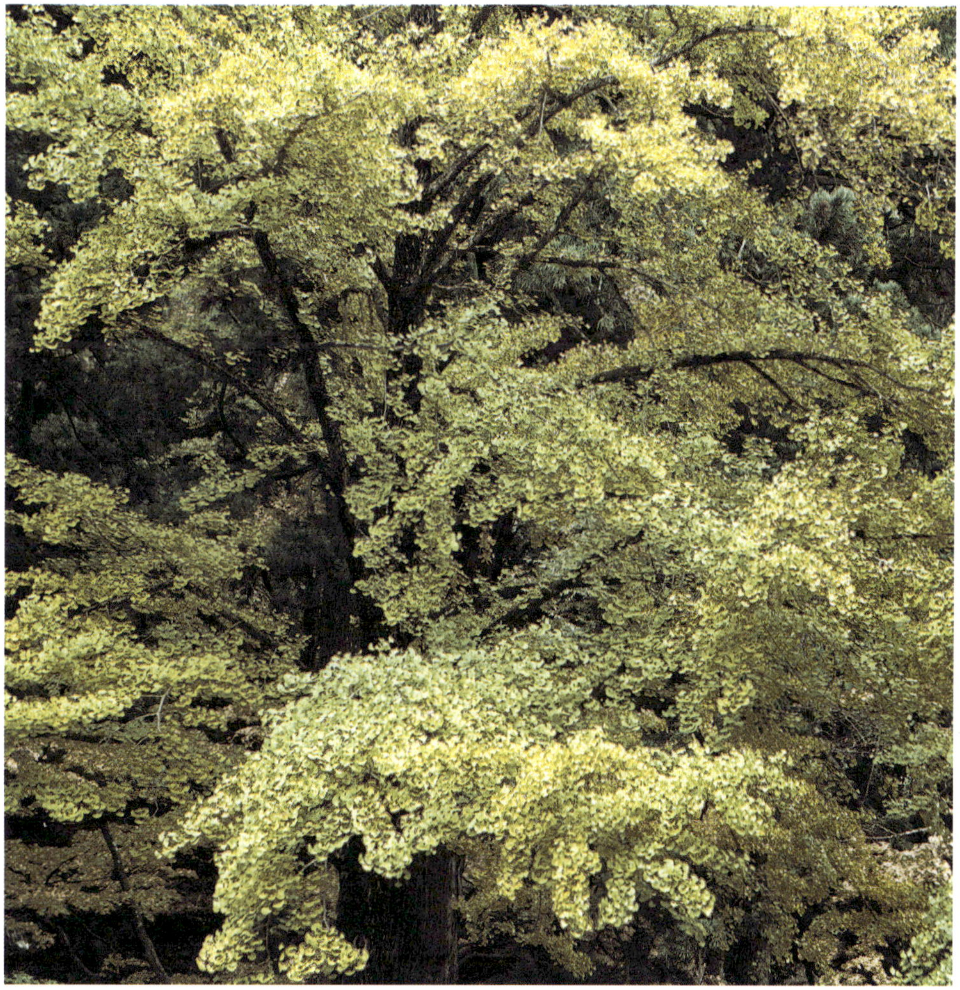

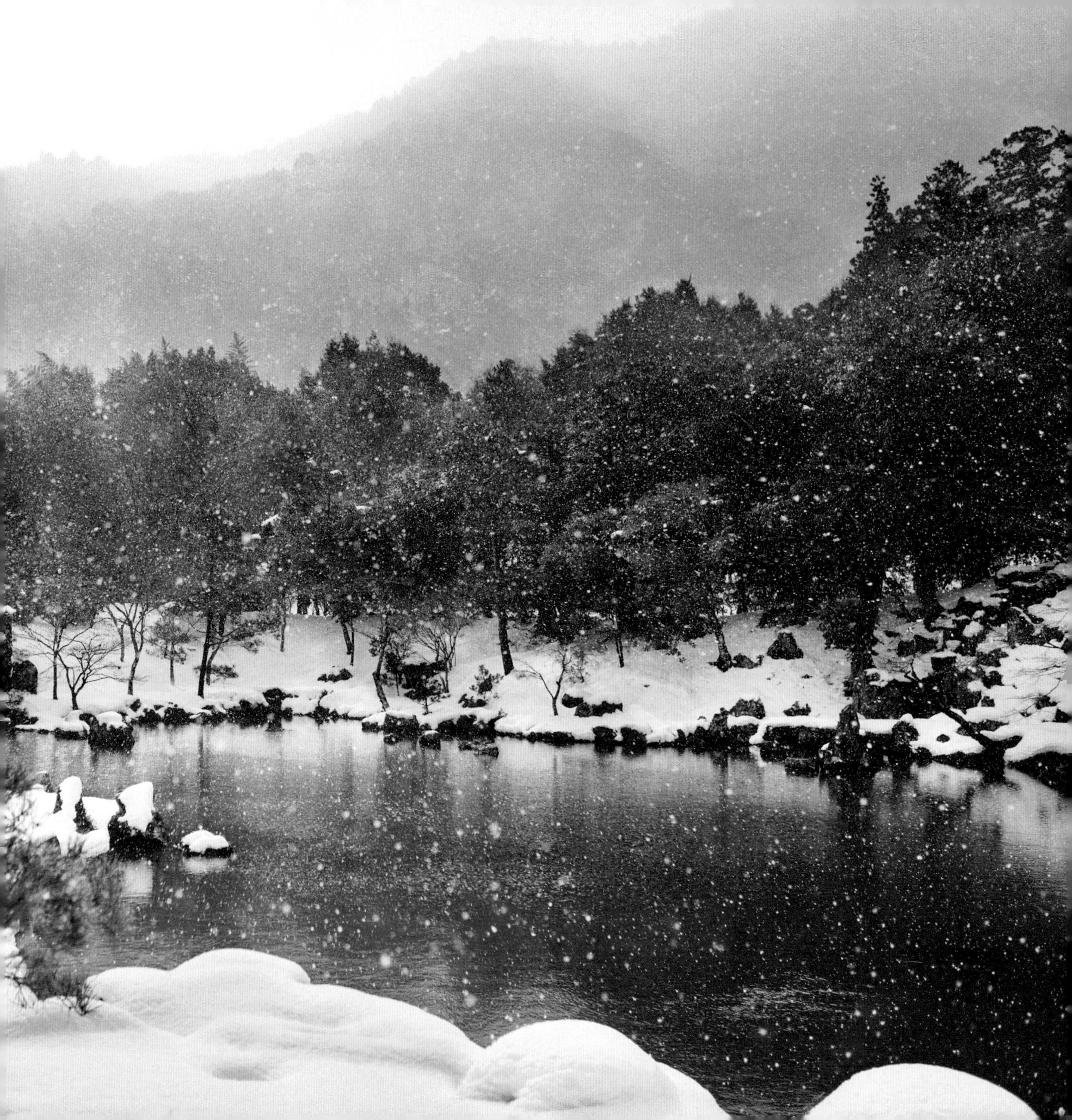

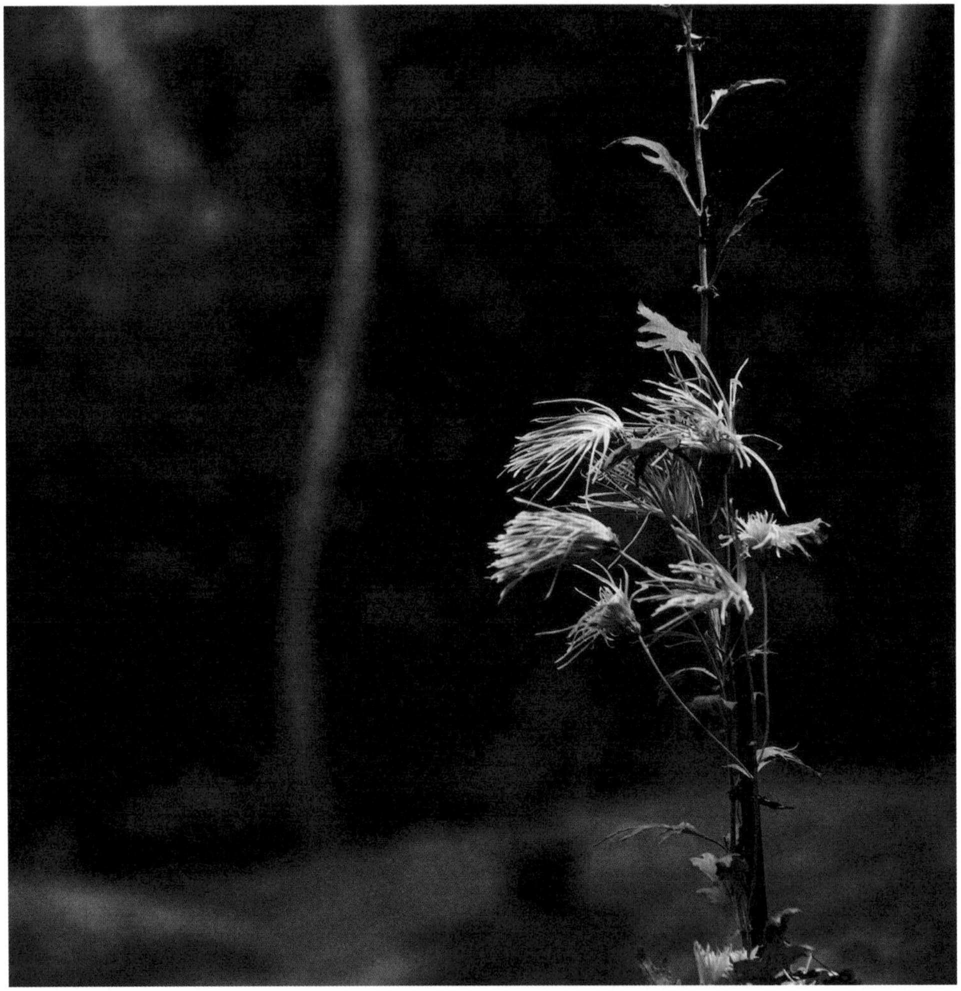

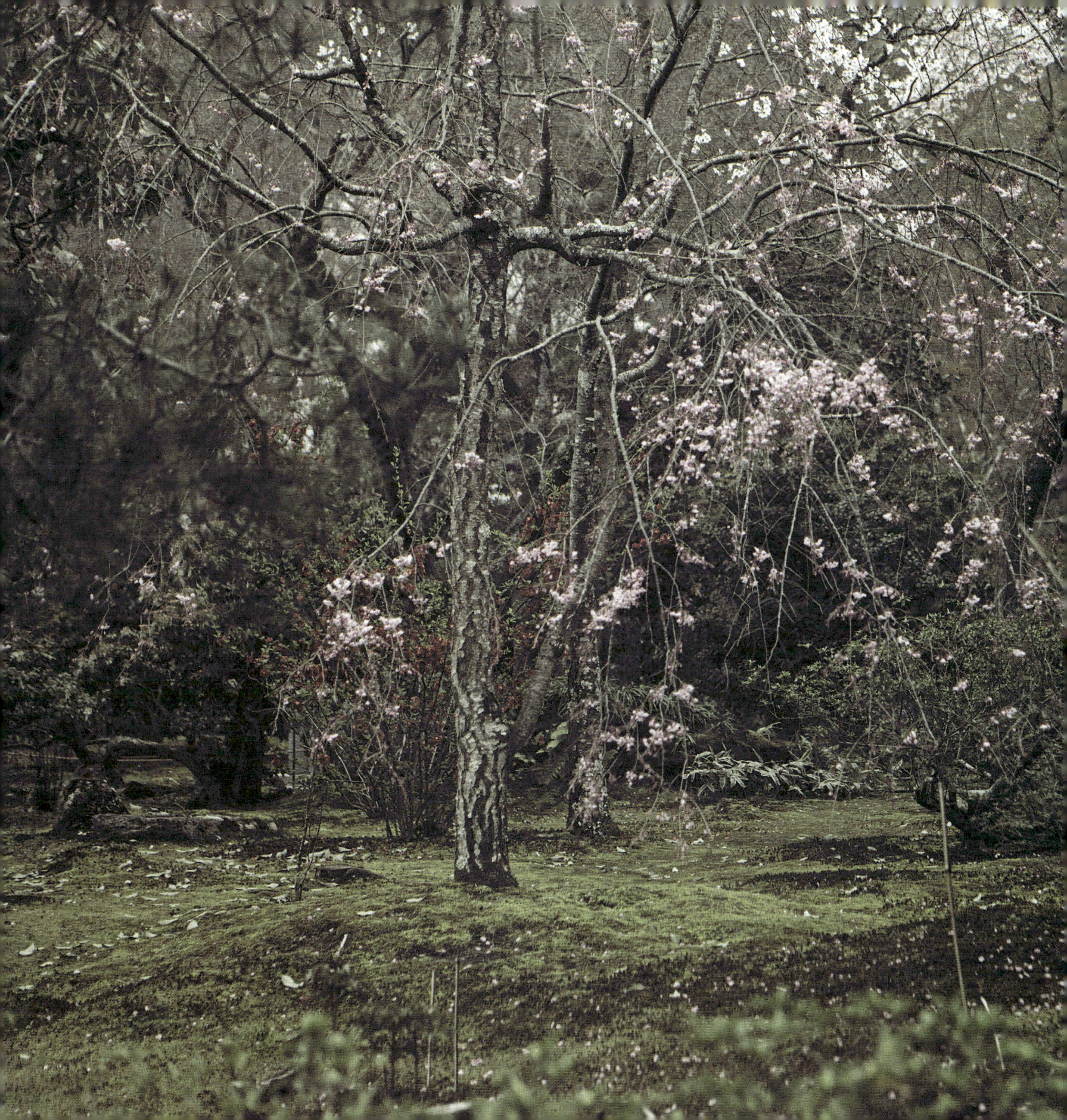

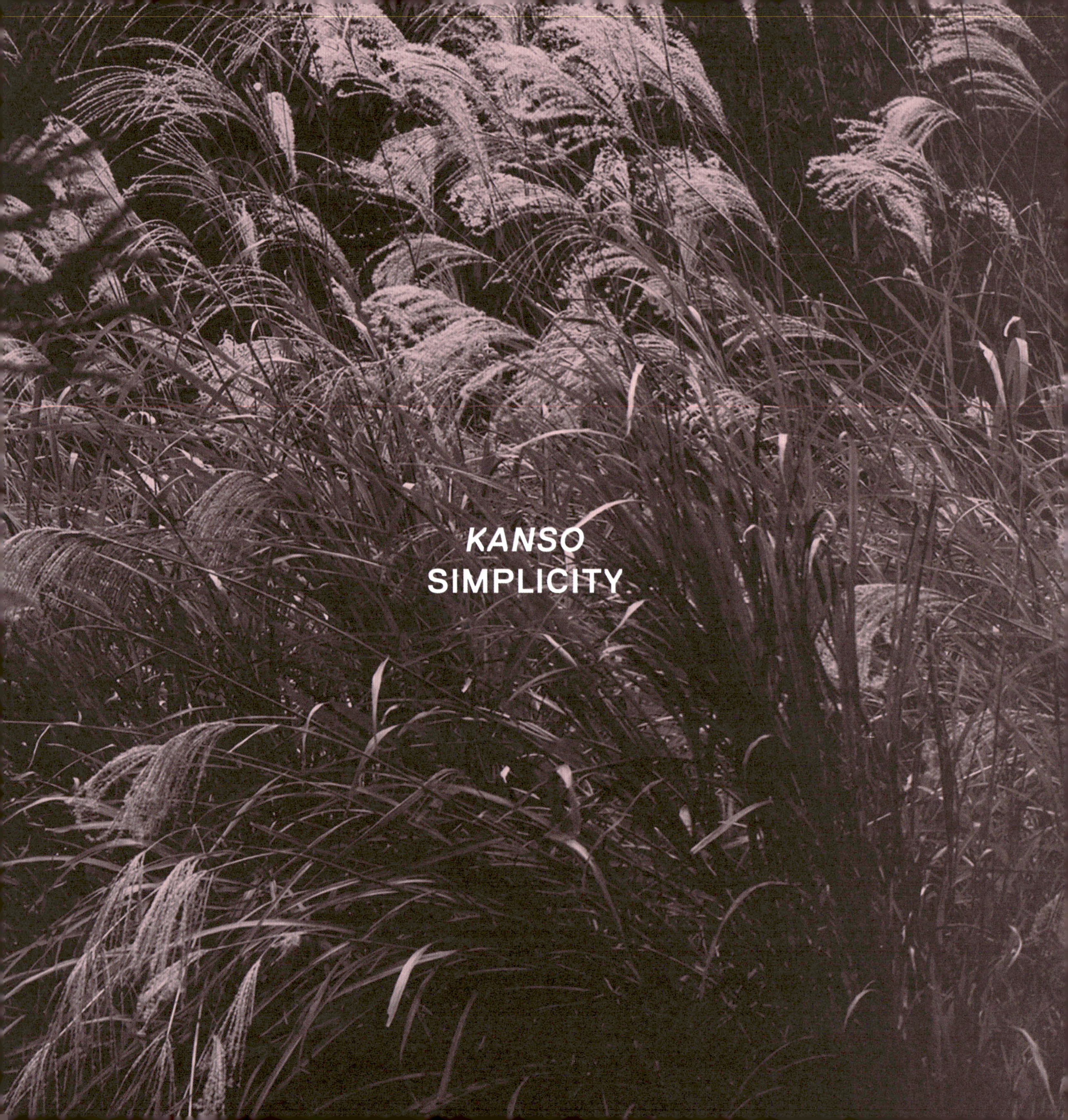
KANSO
SIMPLICITY

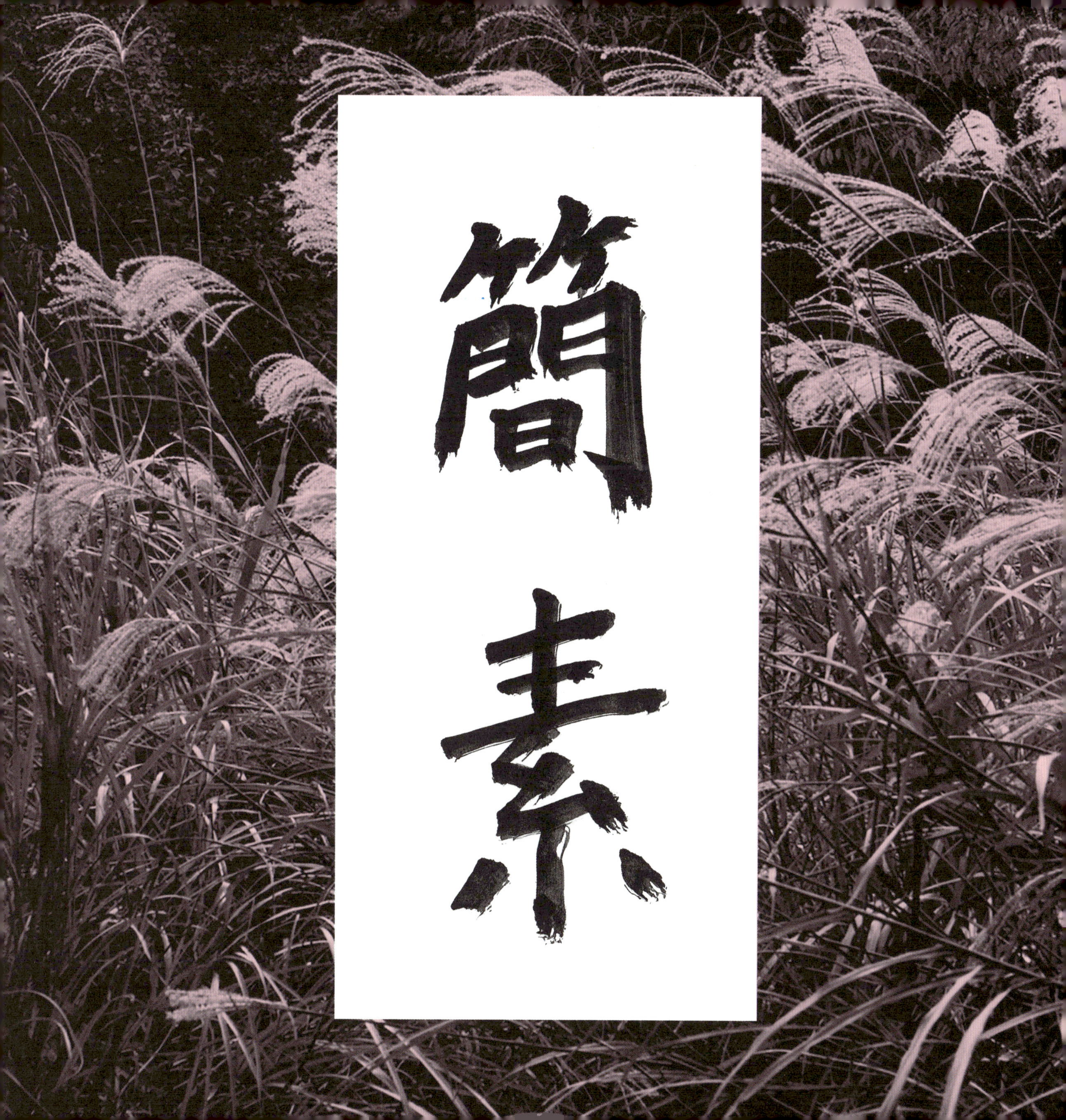
簡素

THE ZEN ARTS DO NOT SERVE ANY USEFUL PURPOSE, NOR DO THEY SERVE PURE AESTHETIC PLEASURE. RATHER, THEY AIM TO EXERCISE THE MIND, TO BRING IT INTO CONTACT WITH ULTIMATE REALITY.

Suzuki Daisetsu

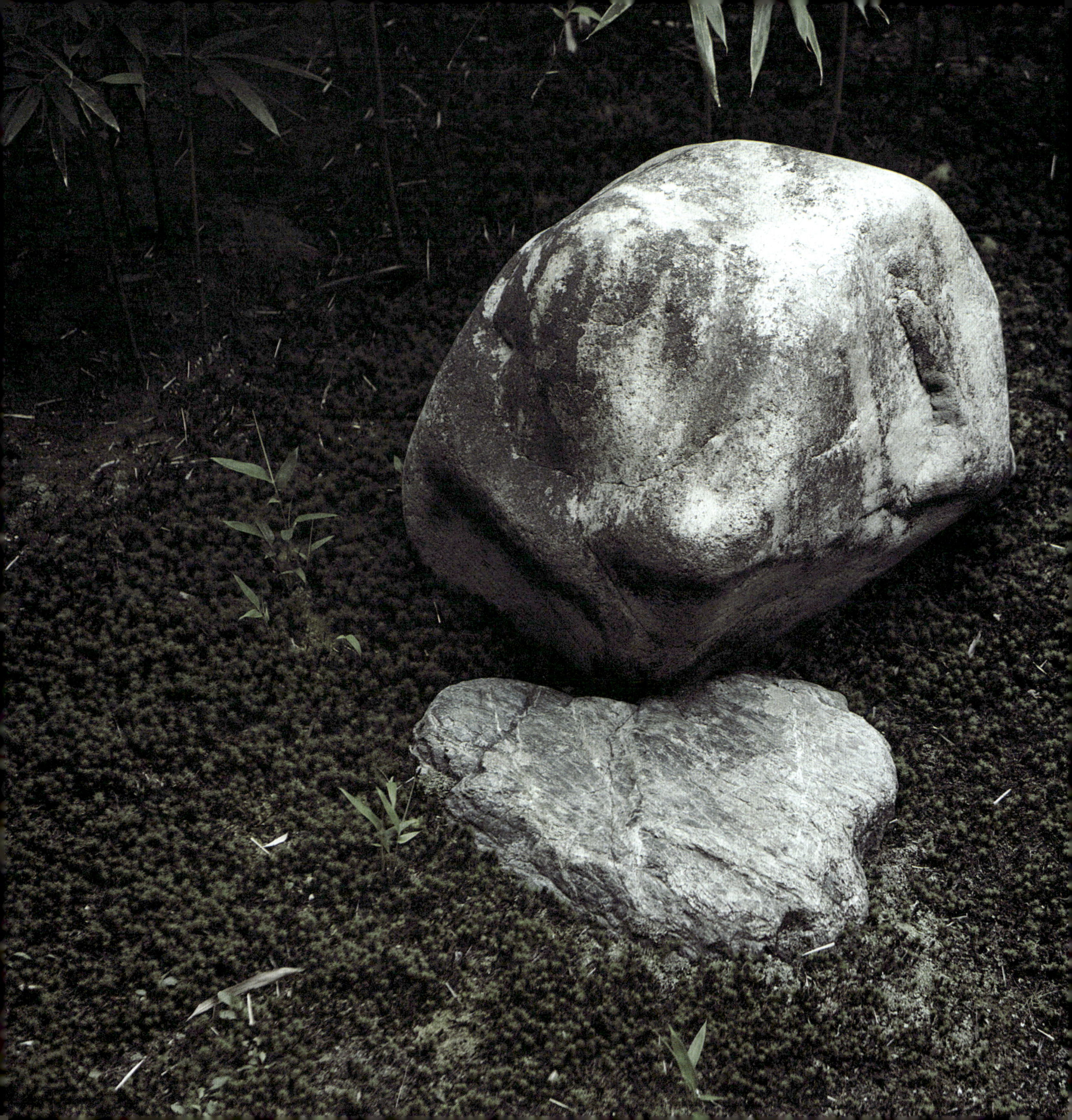

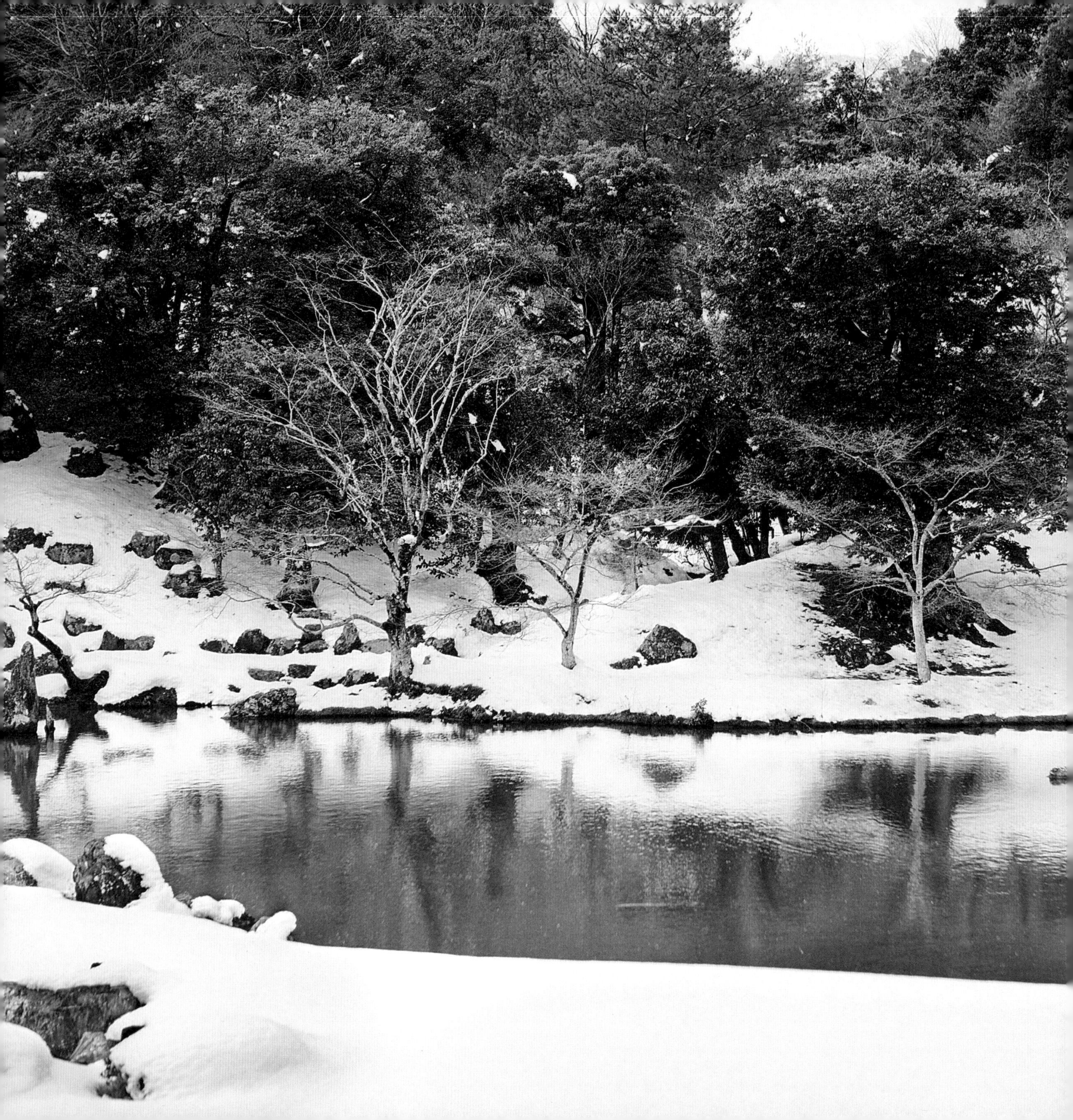

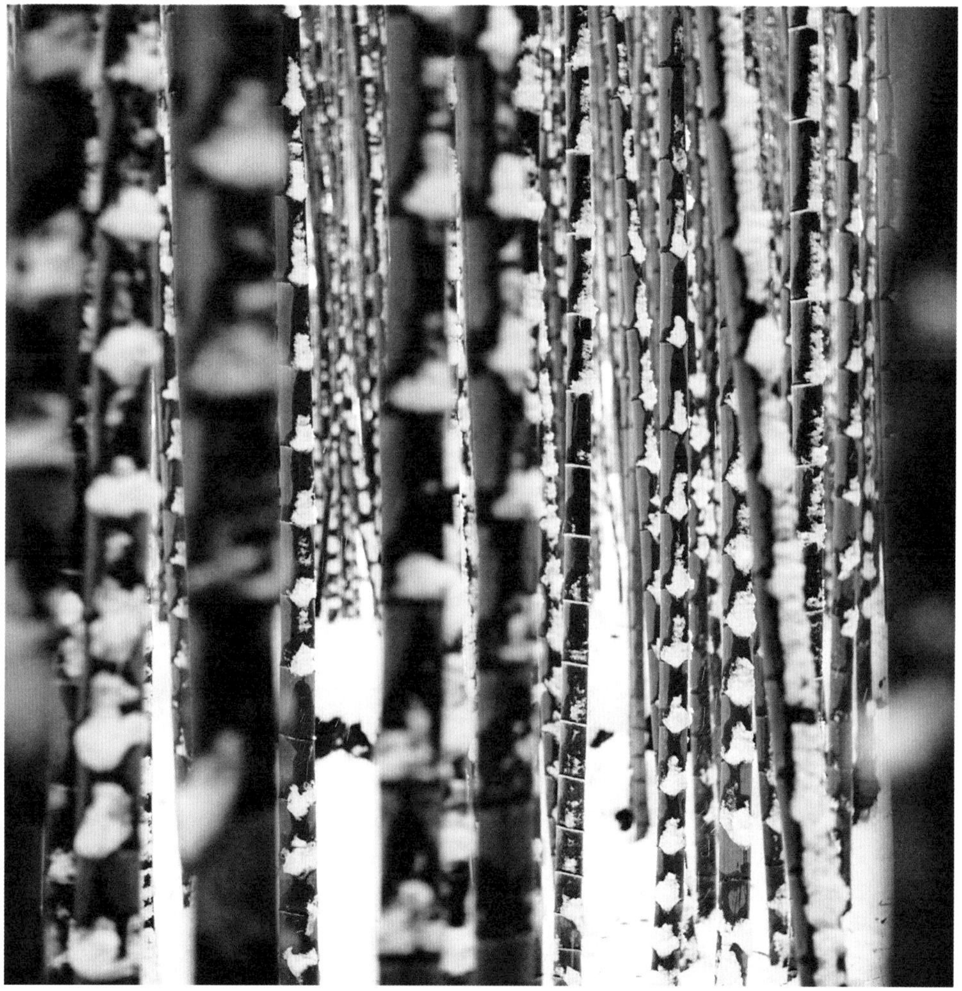

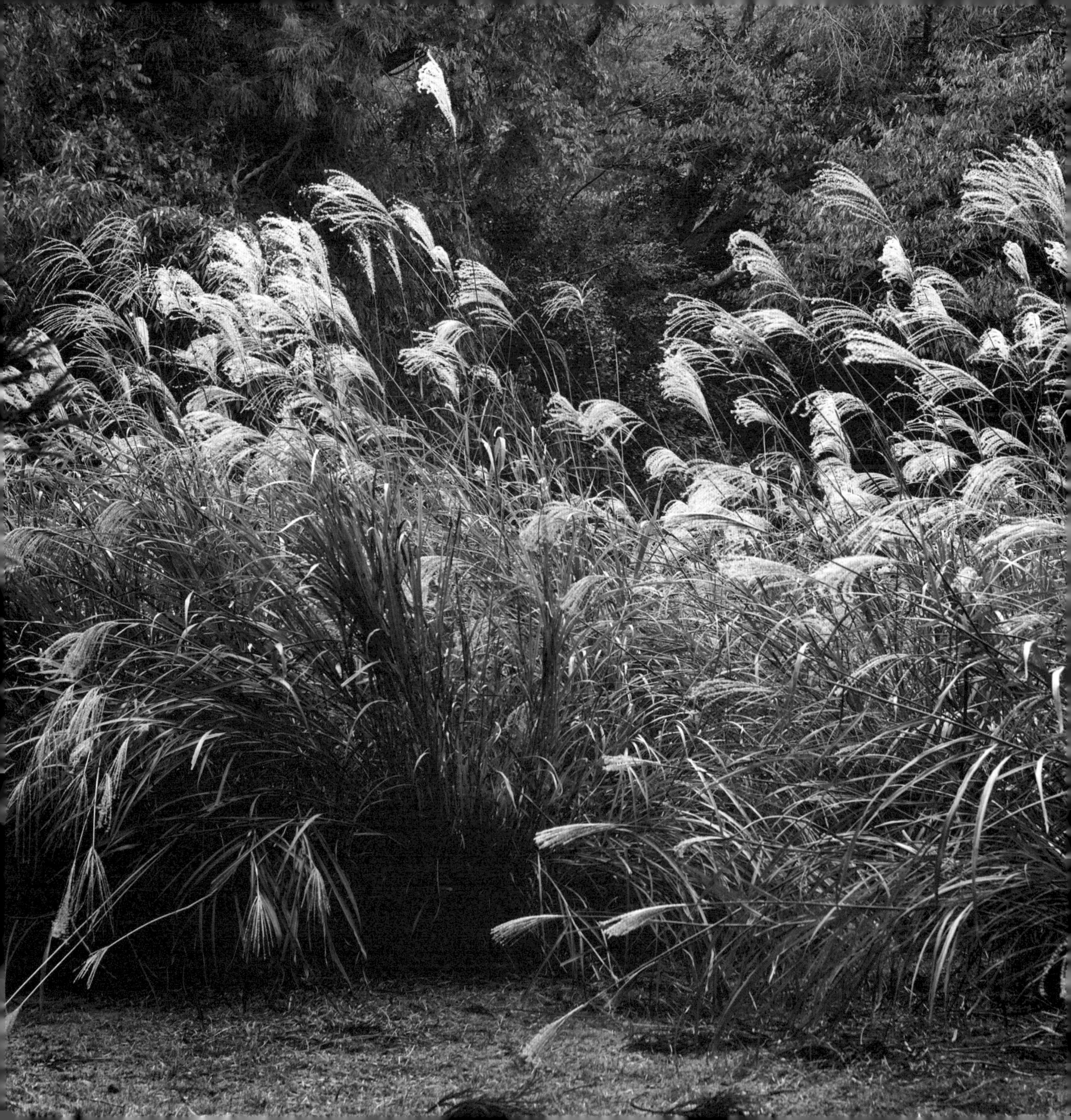

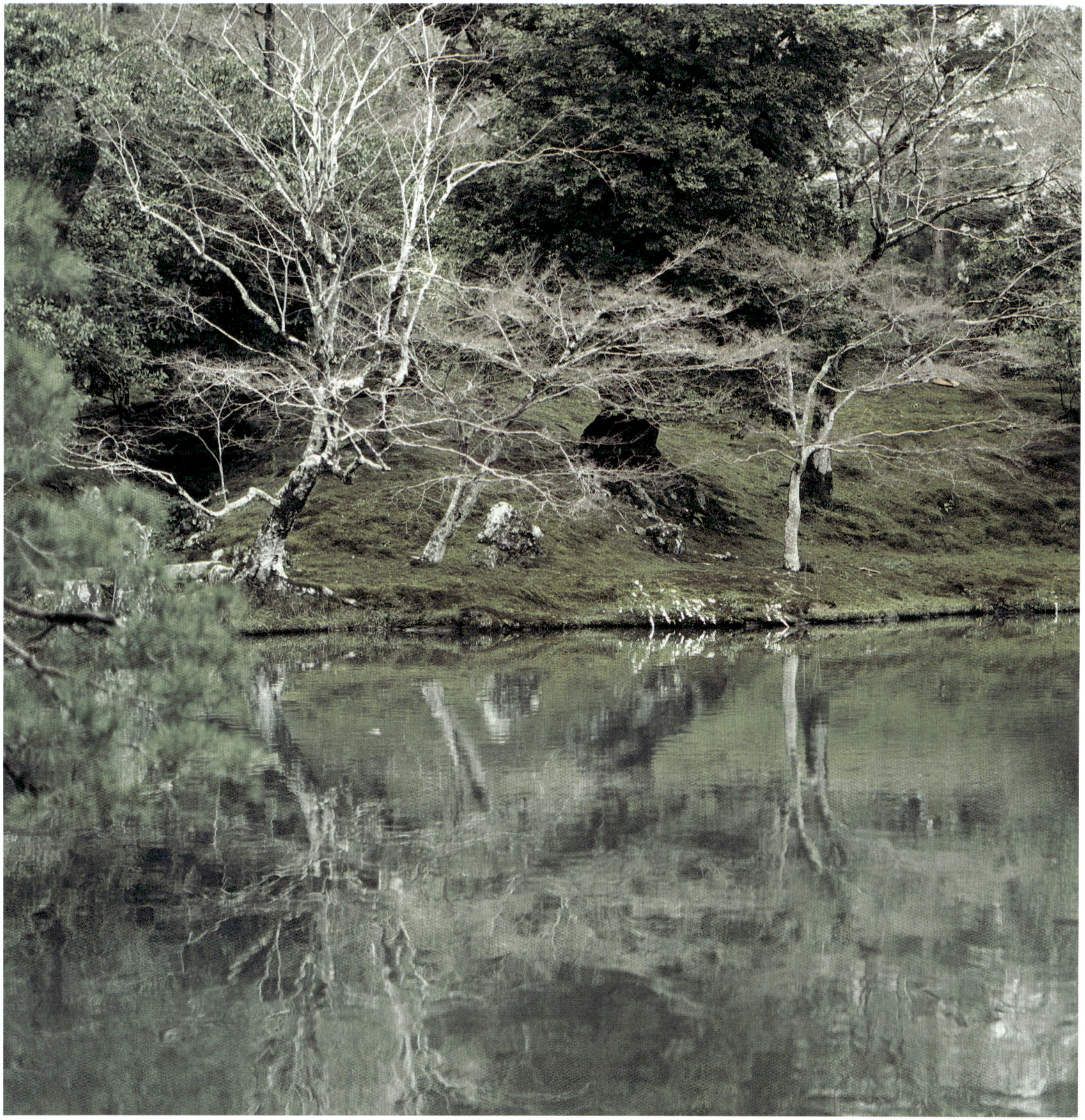

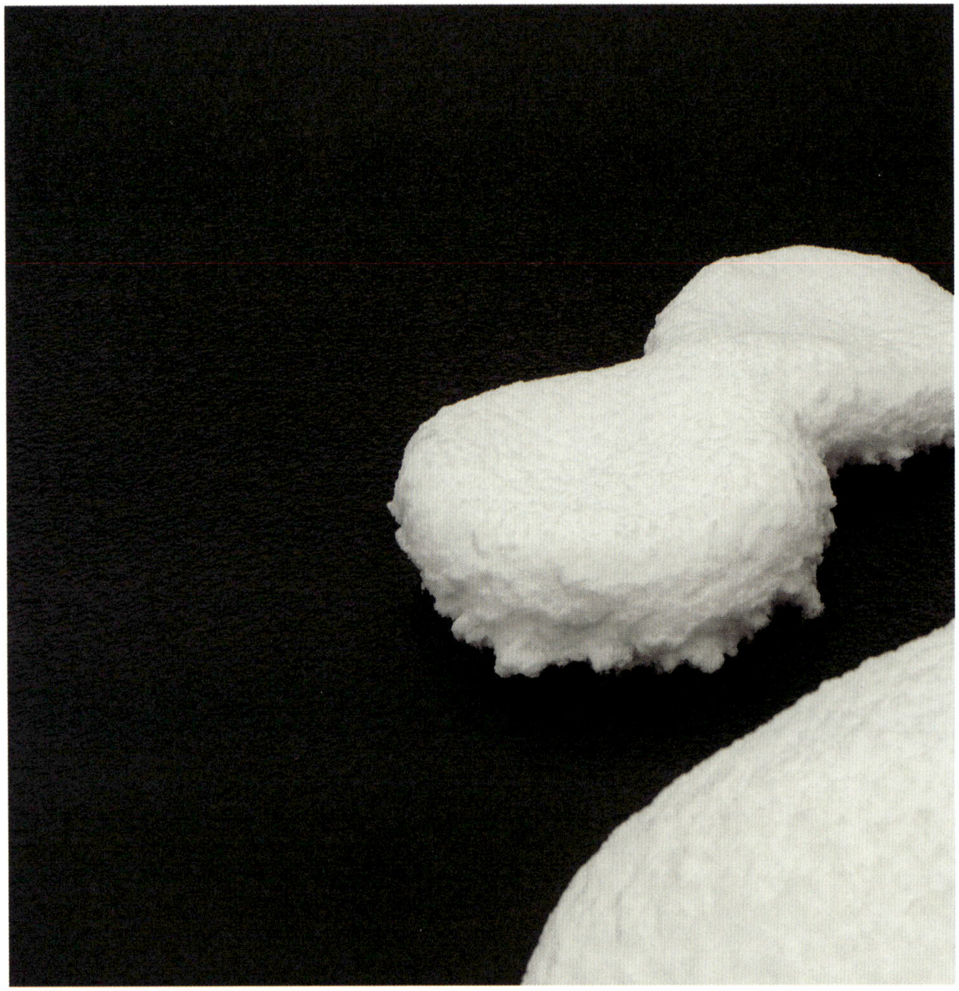

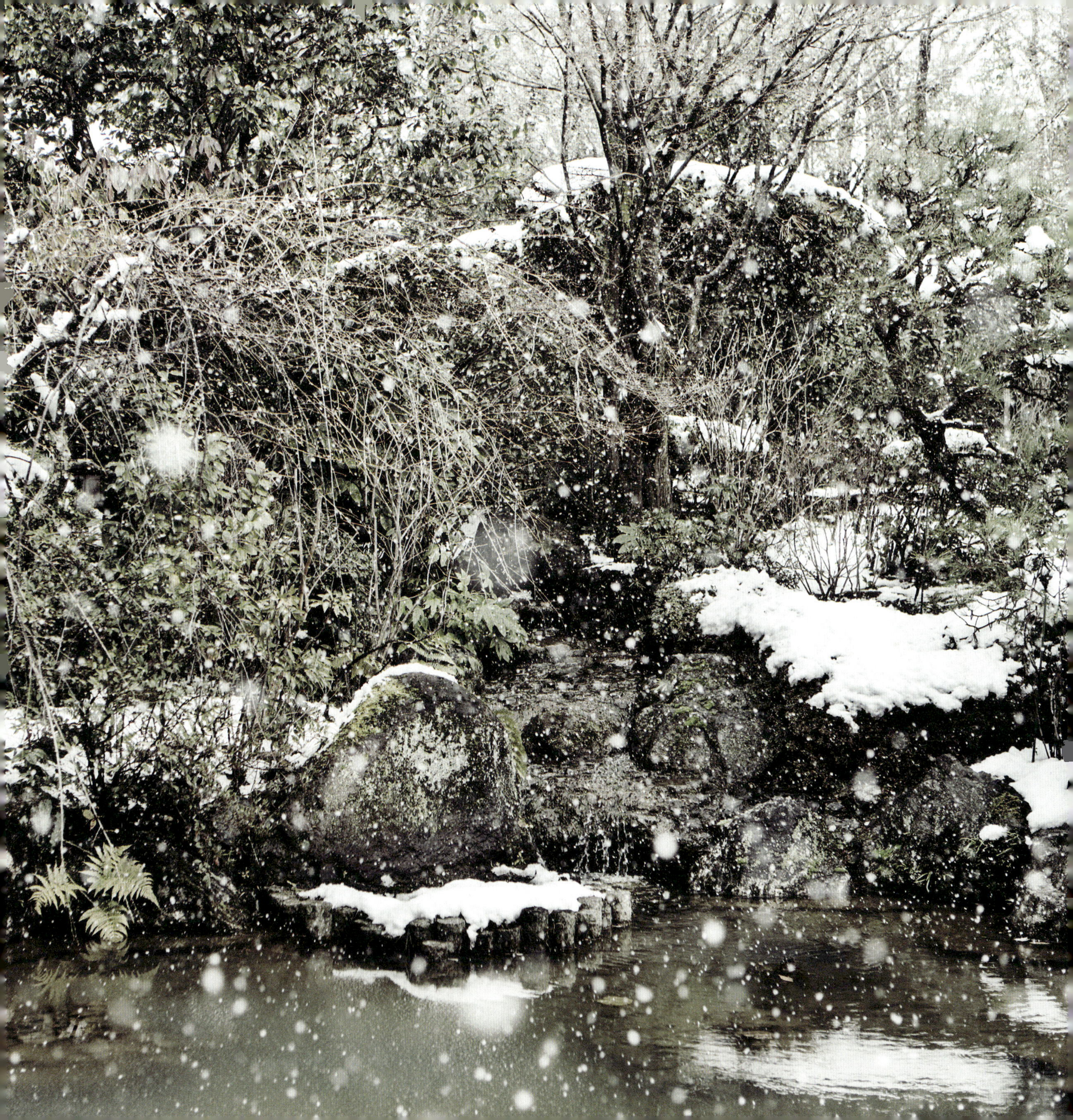

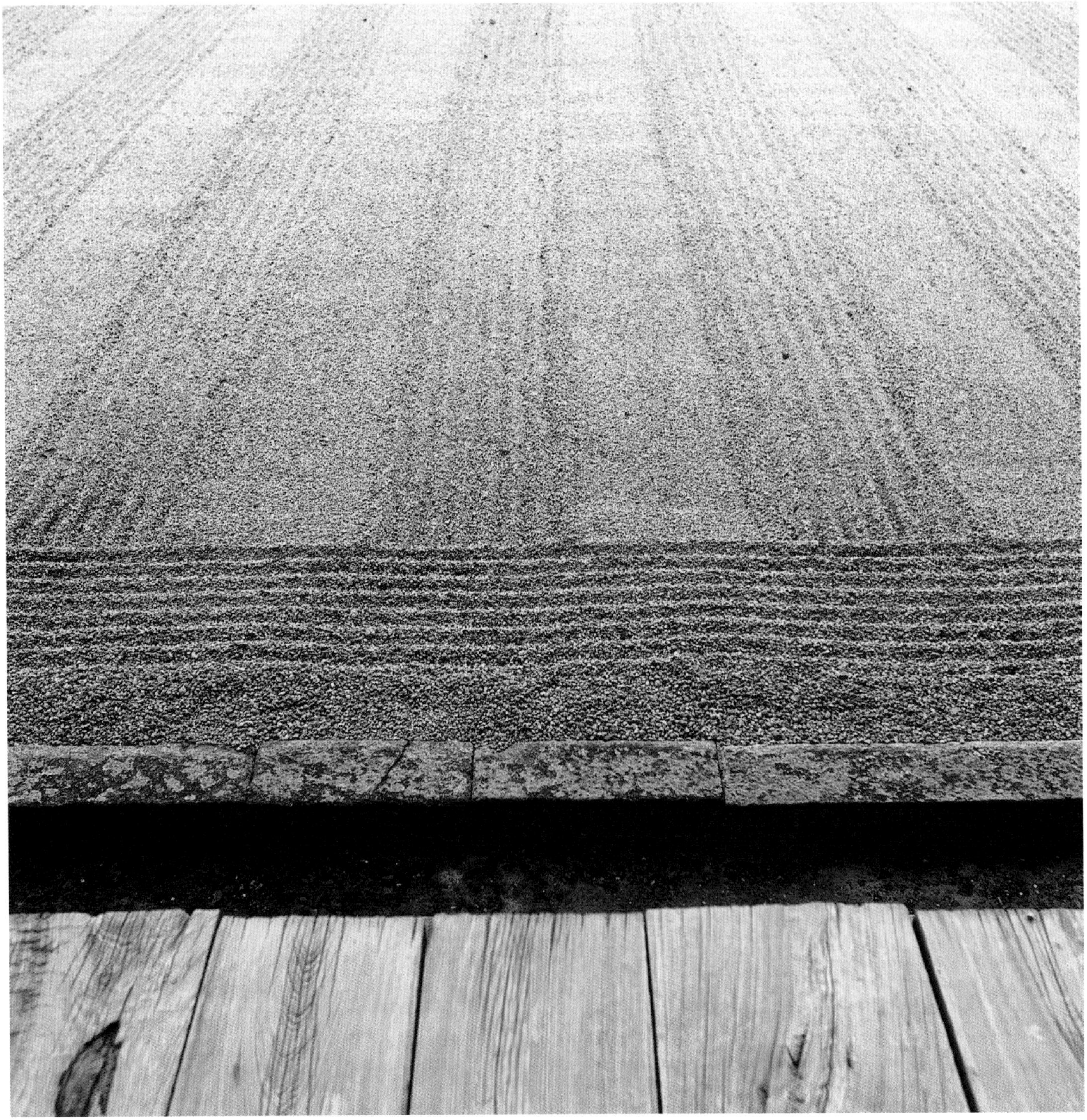

WHAT WAS YOUR FACE
BEFORE YOUR PARENTS WERE BORN?

Zen Koan

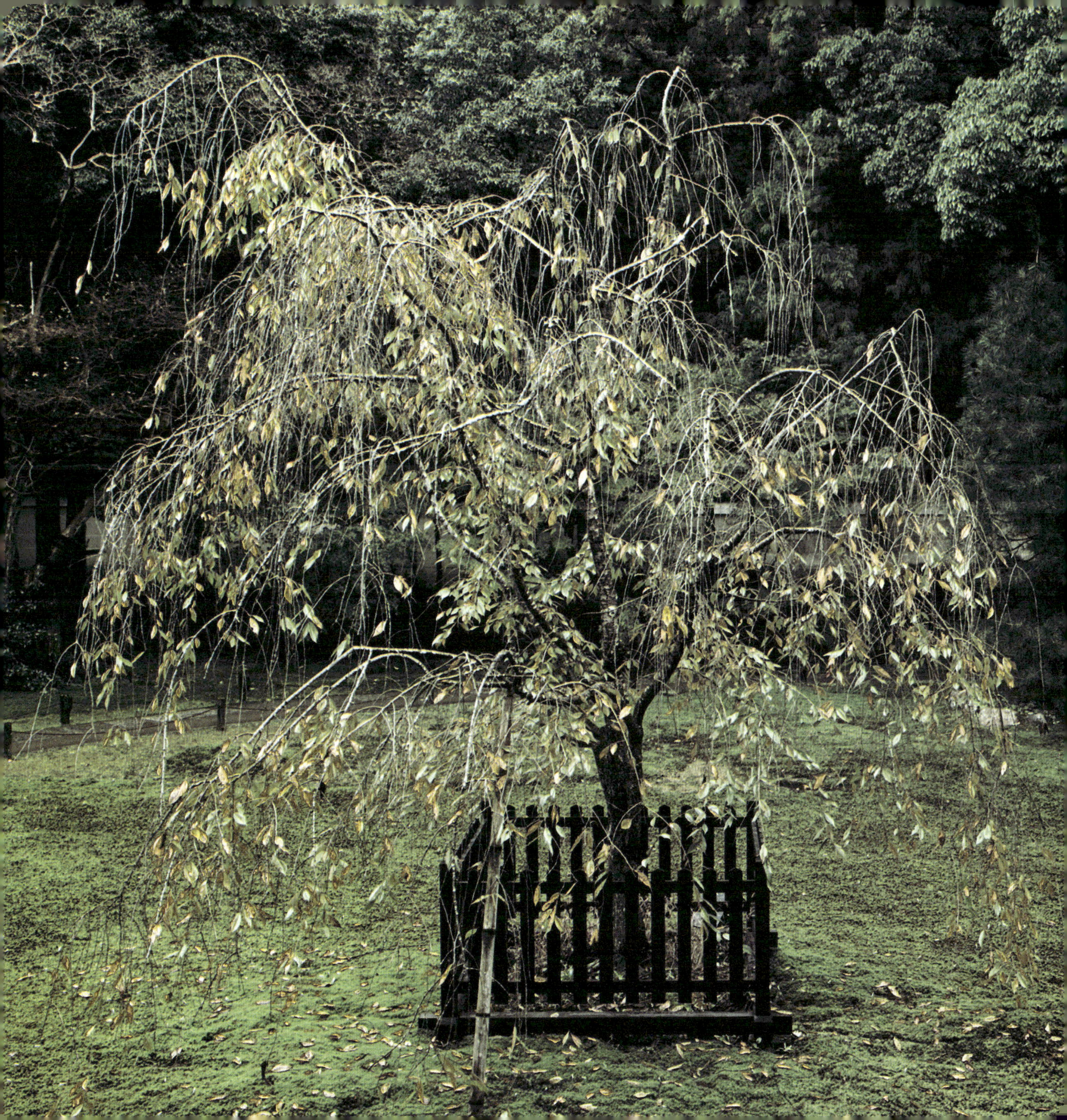

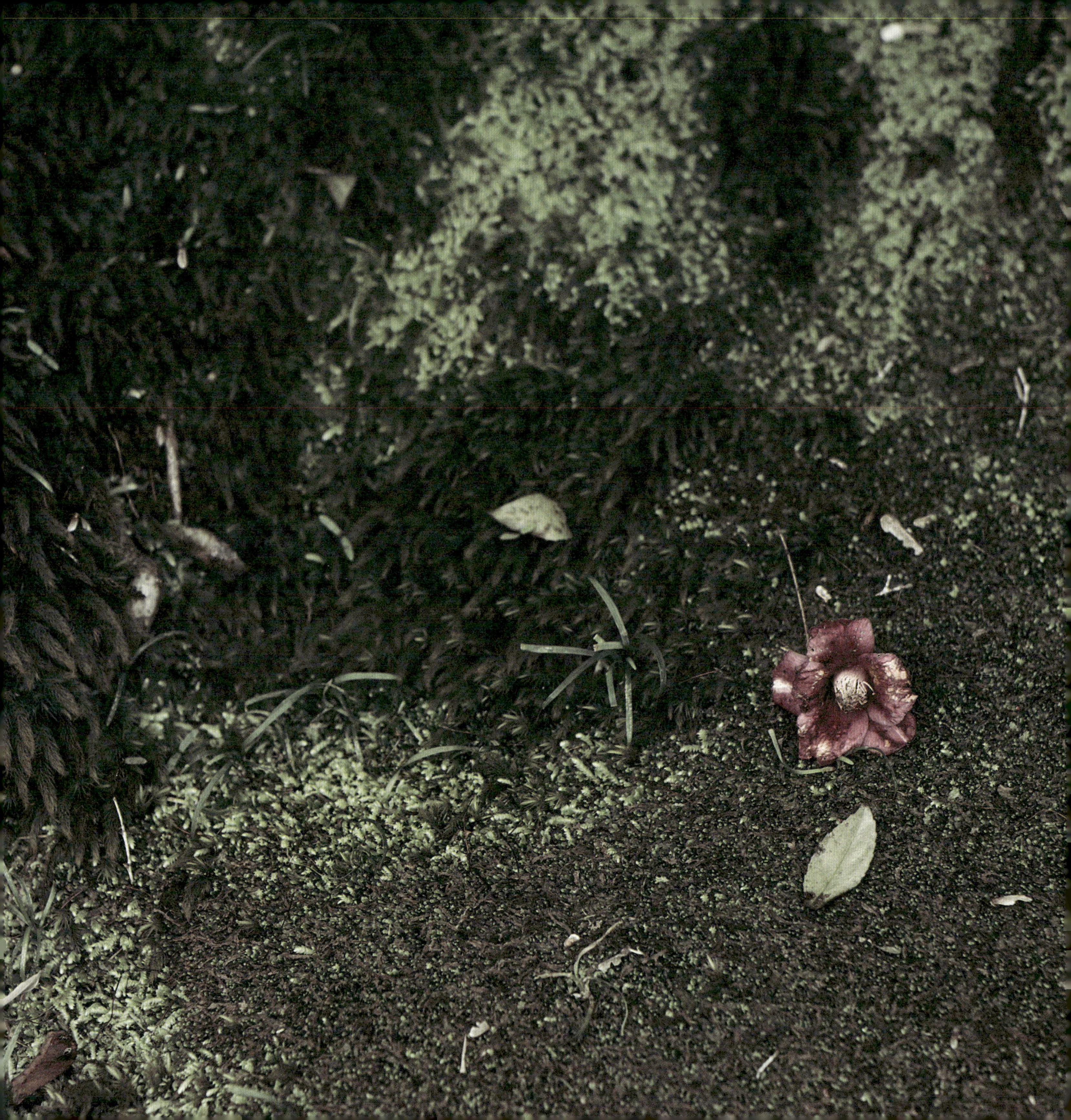

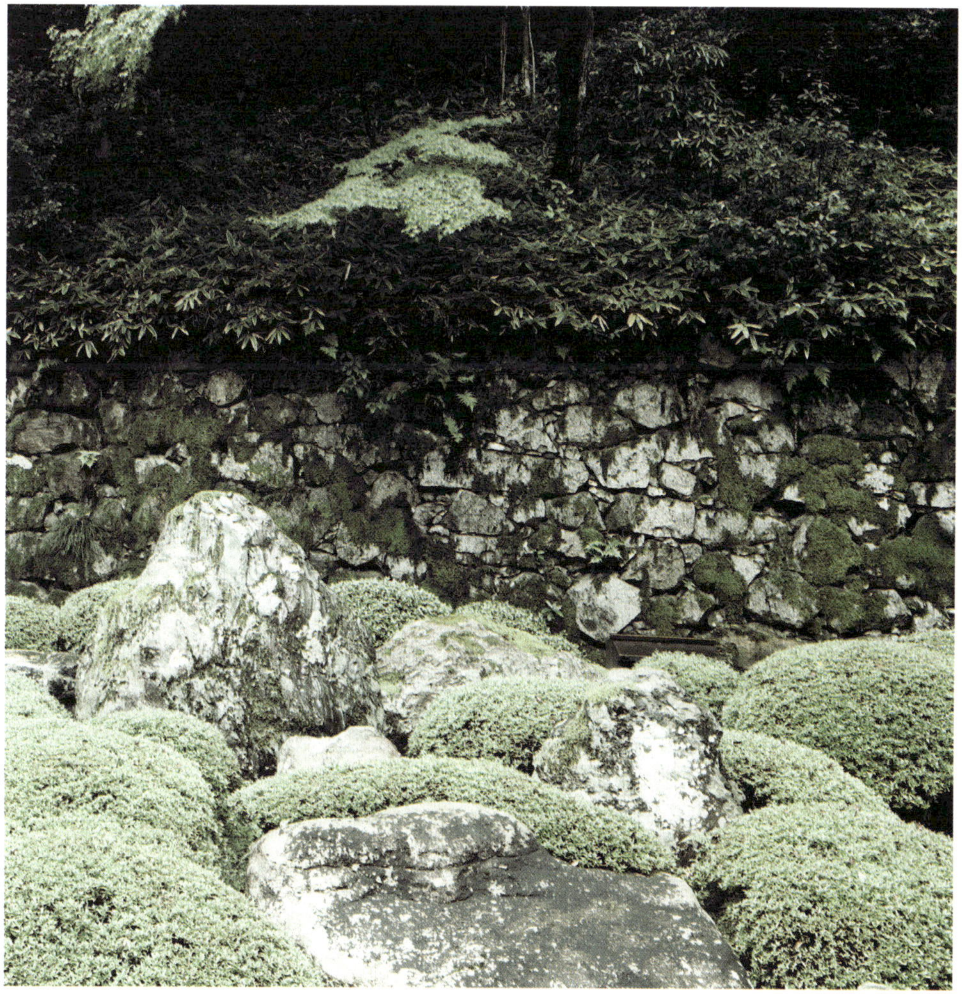

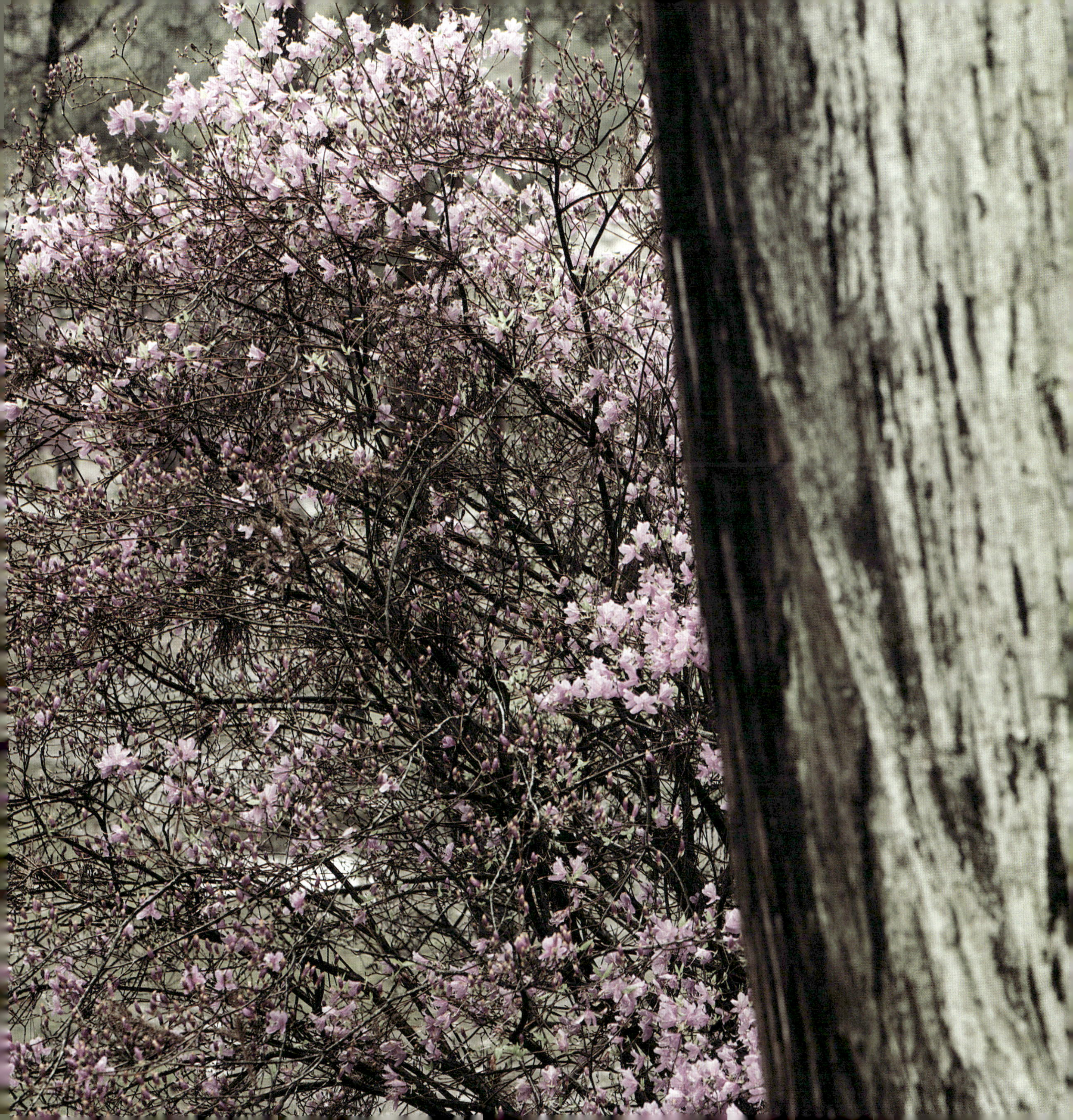

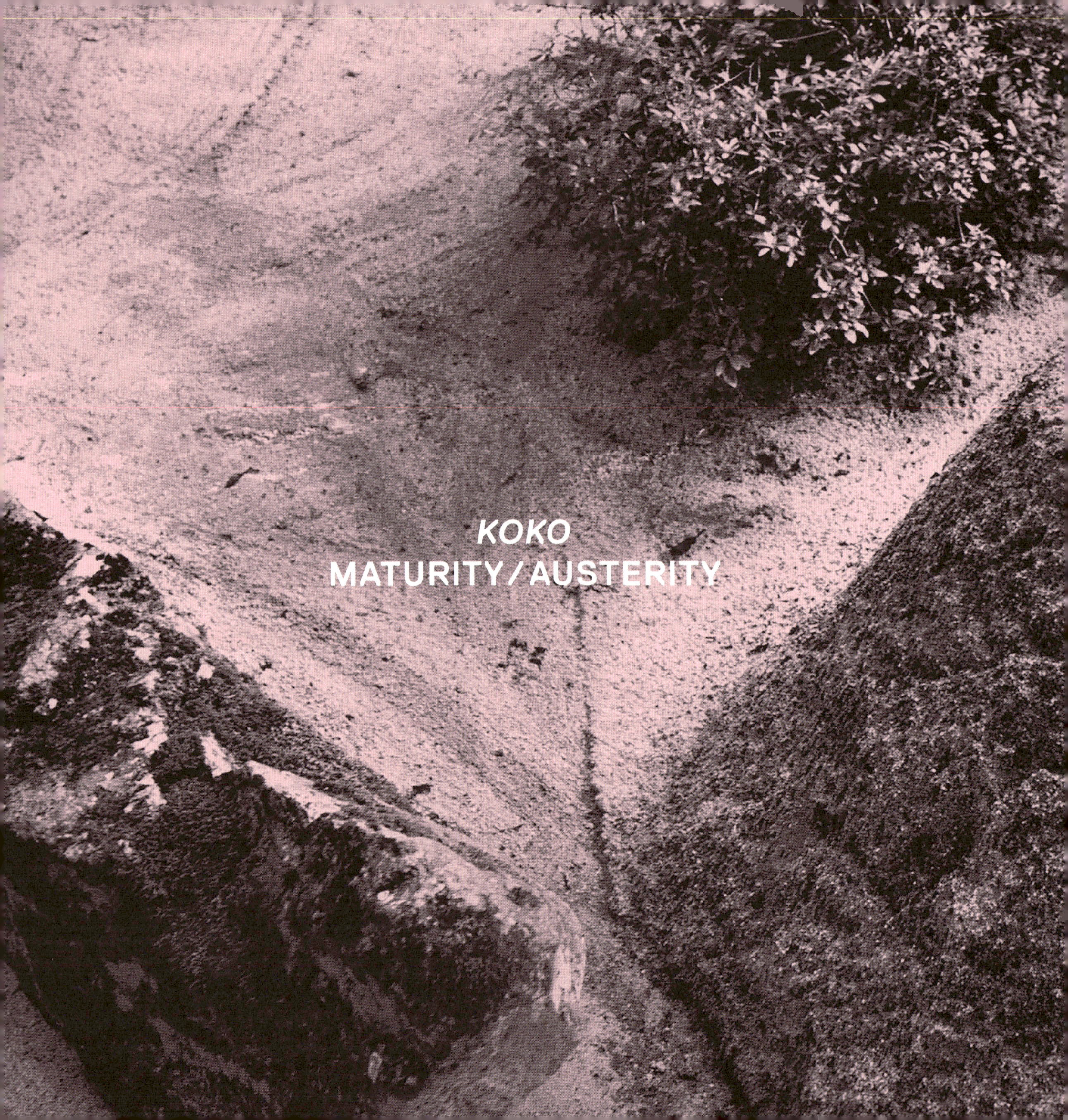

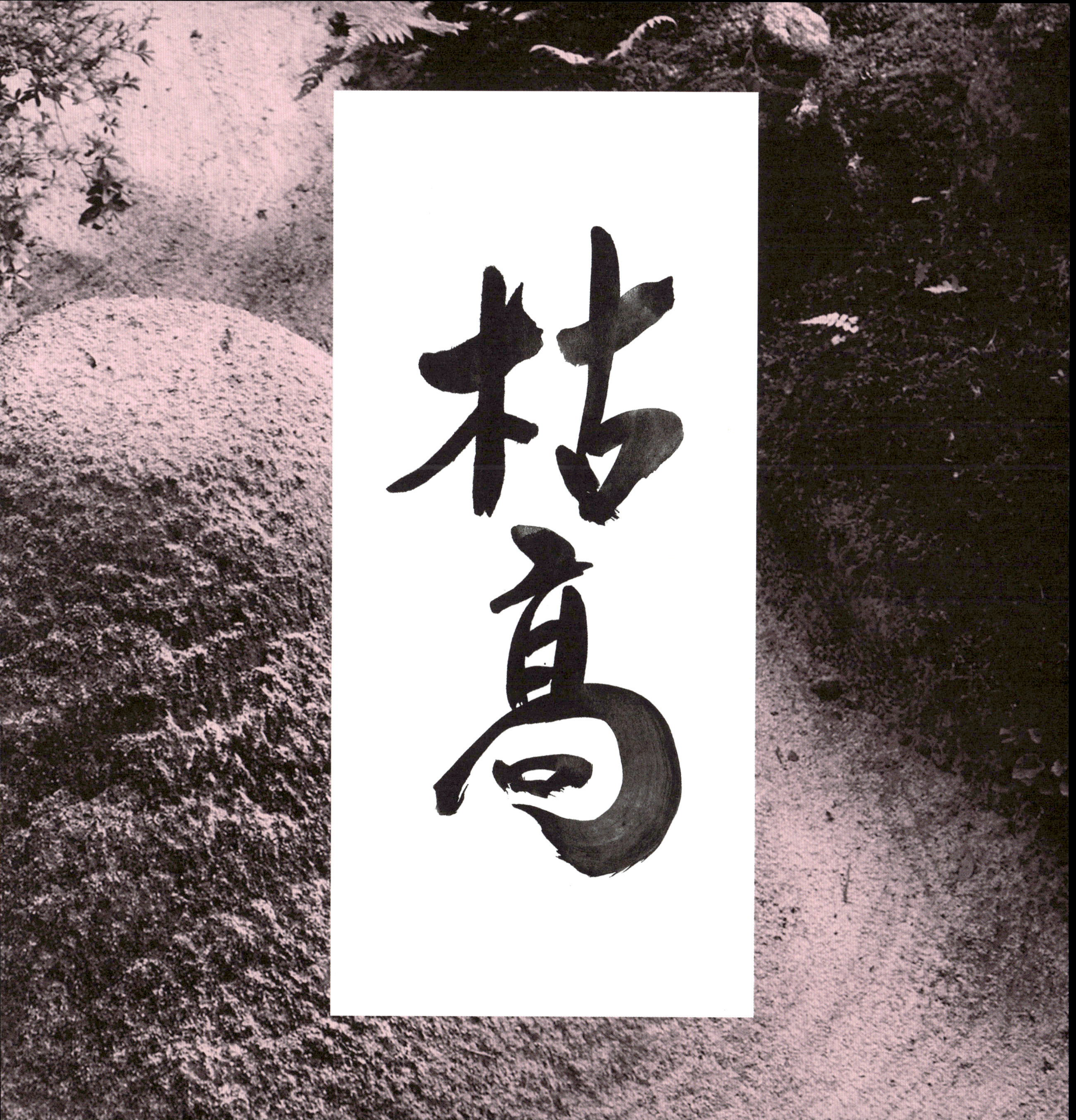
枯高

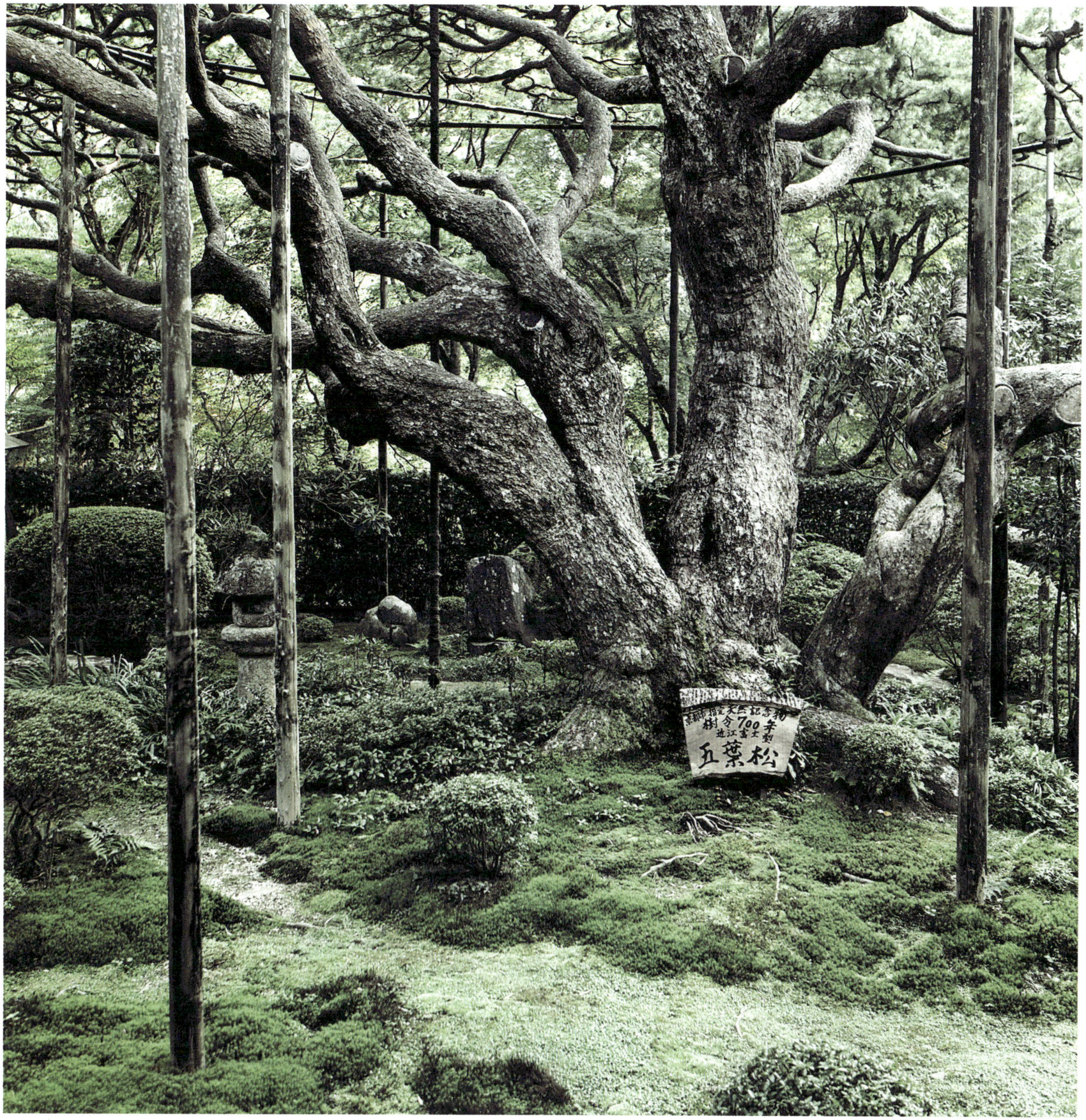
京都市指定天然記念物
樹令700年
近江富士型
五葉松

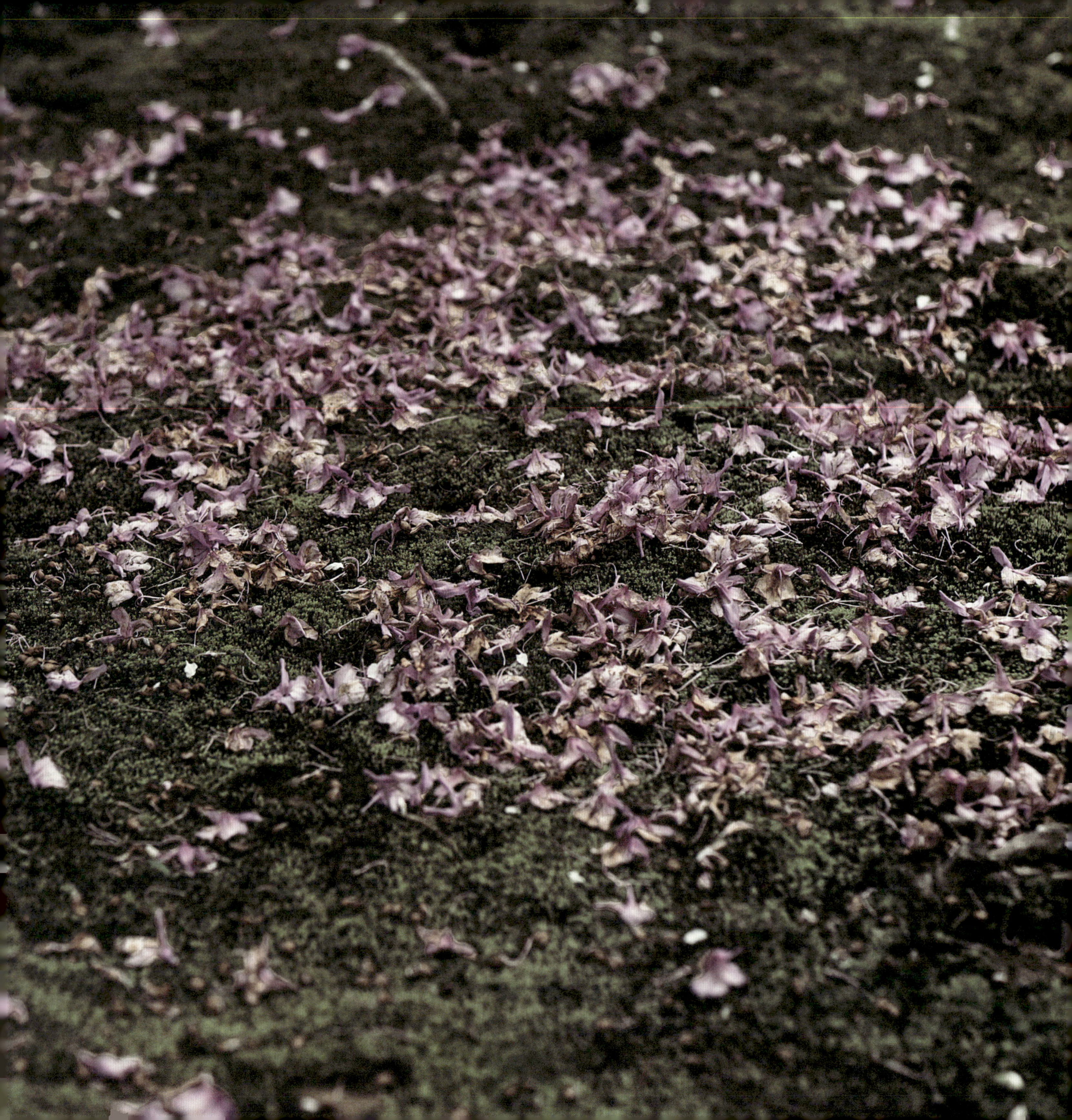

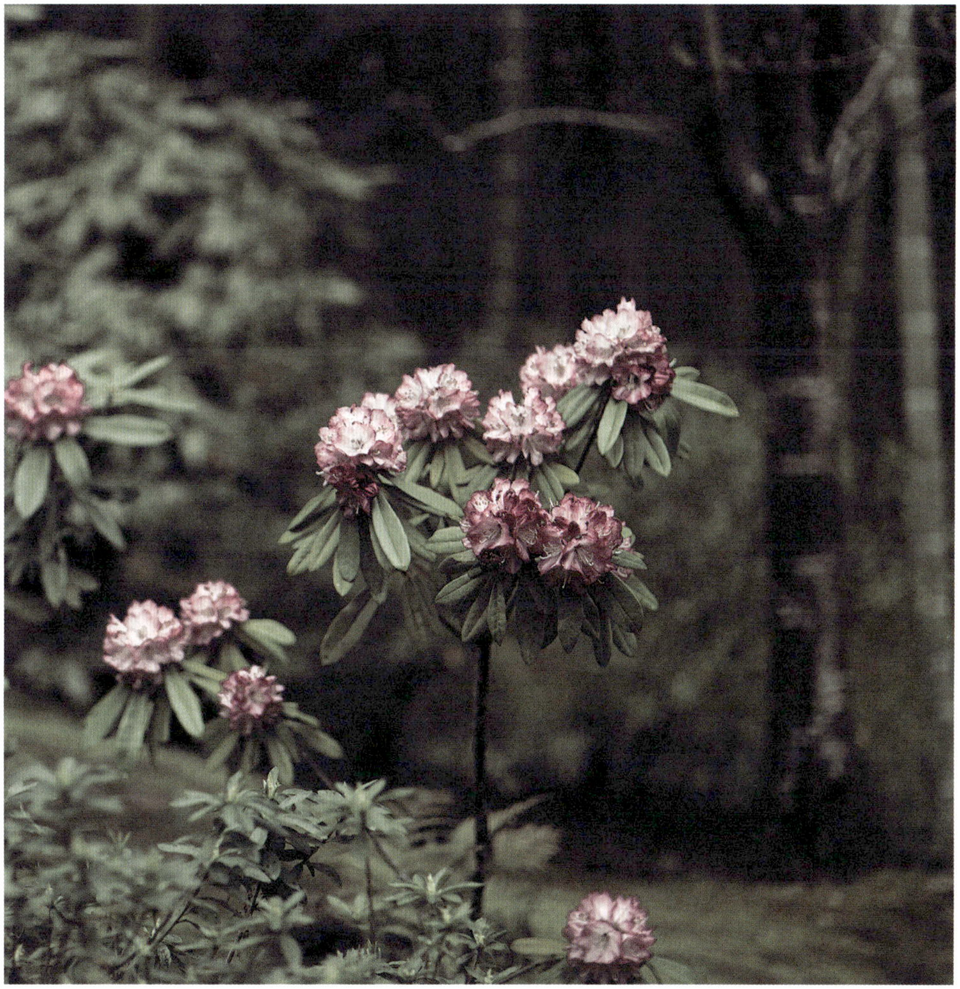

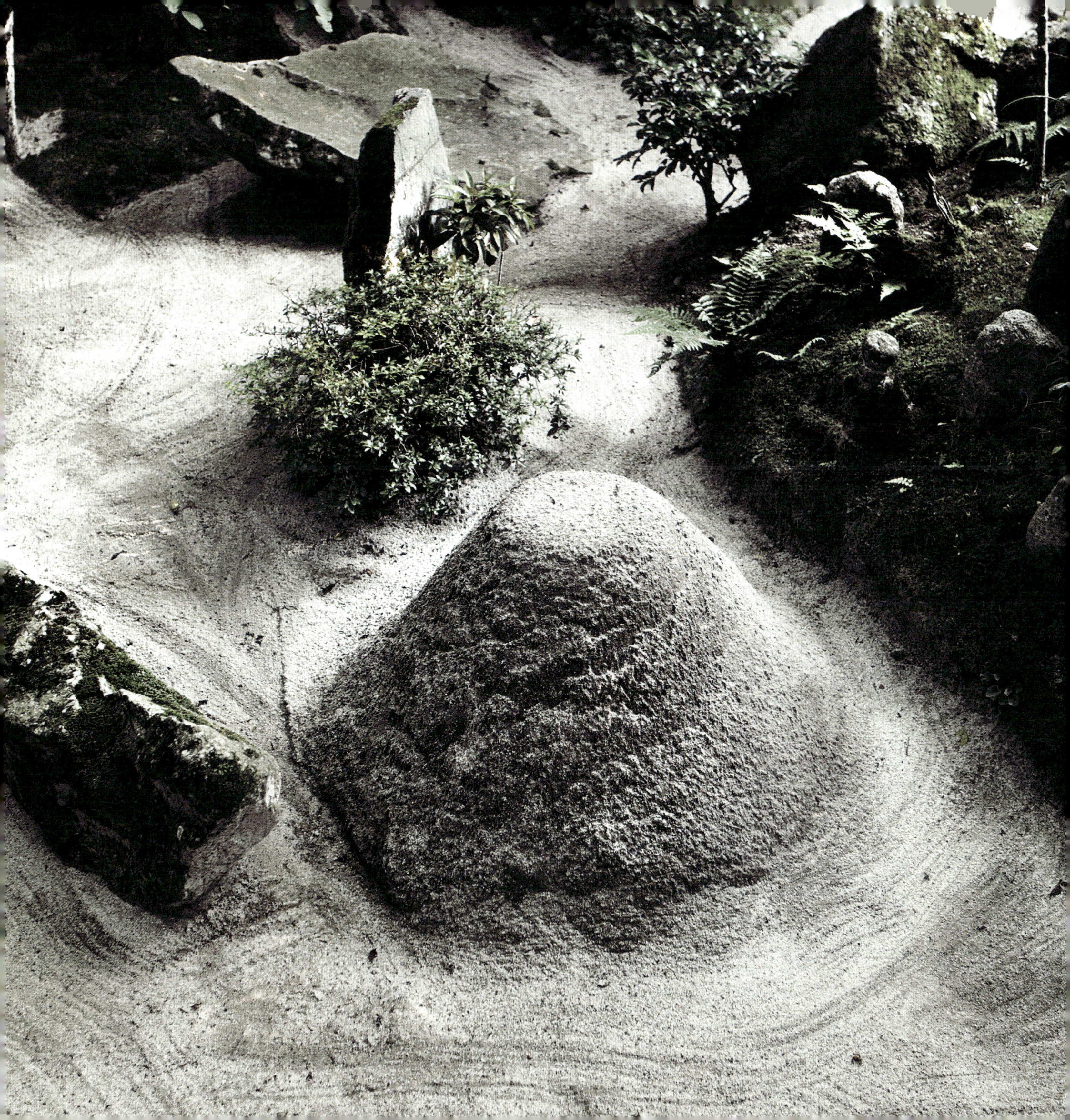

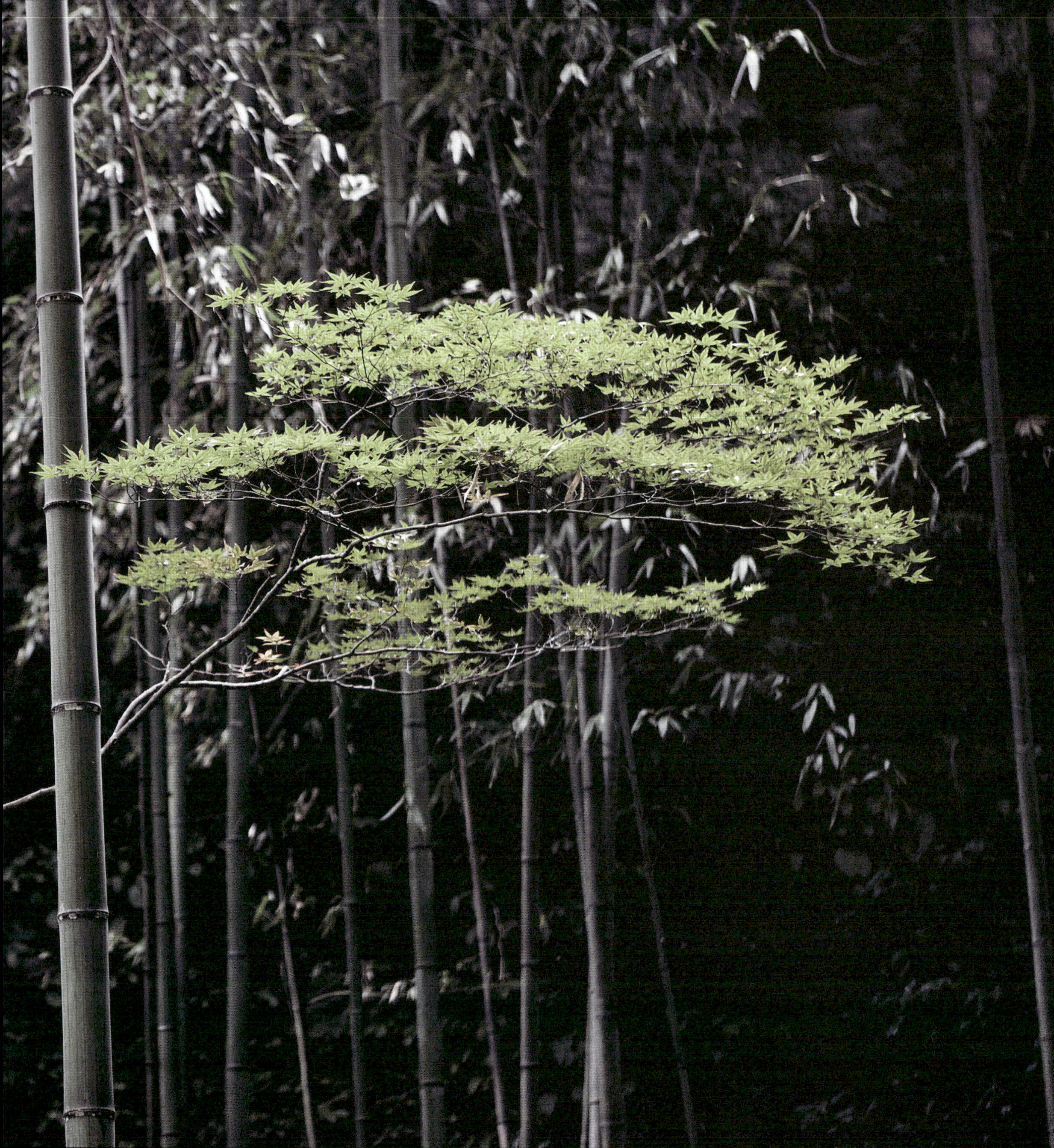

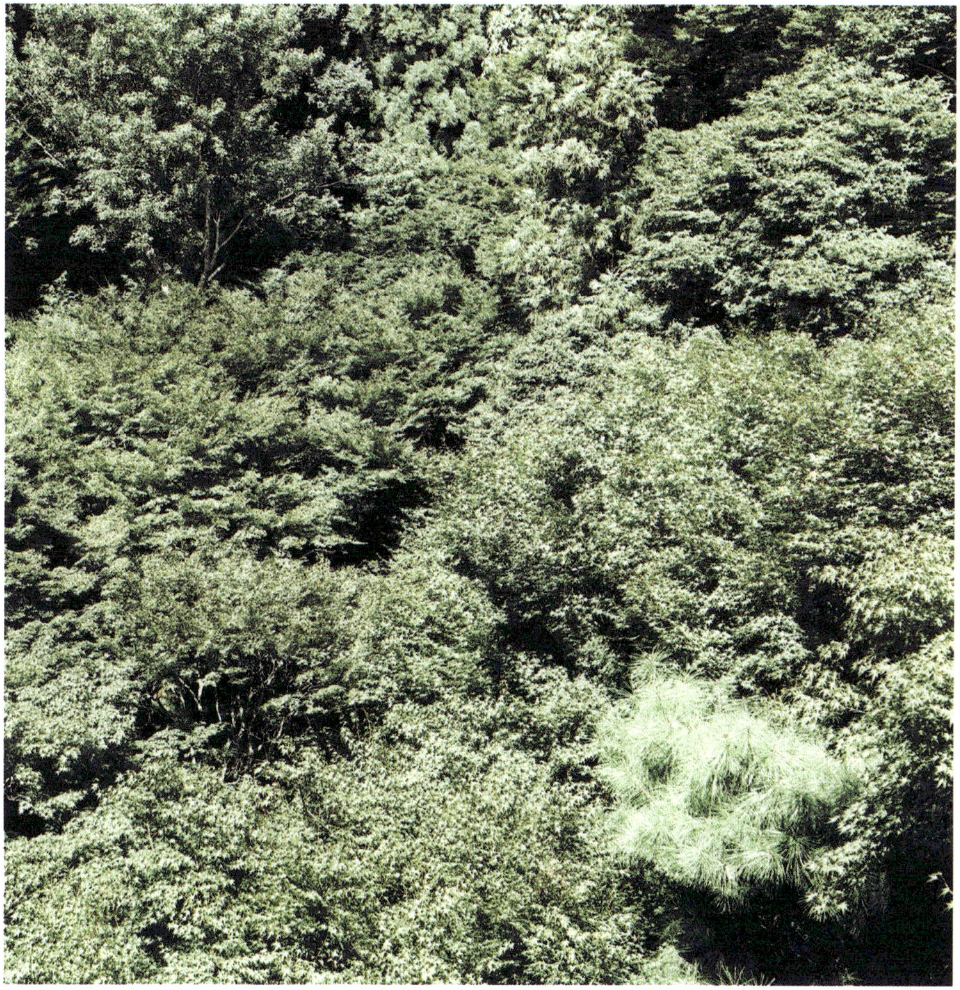

SEEING WITHOUT SEEING.
HEARING WITHOUT HEARING.
FEELING WITHOUT FEELING.
THINKING WITHOUT THINKING.

Unknown

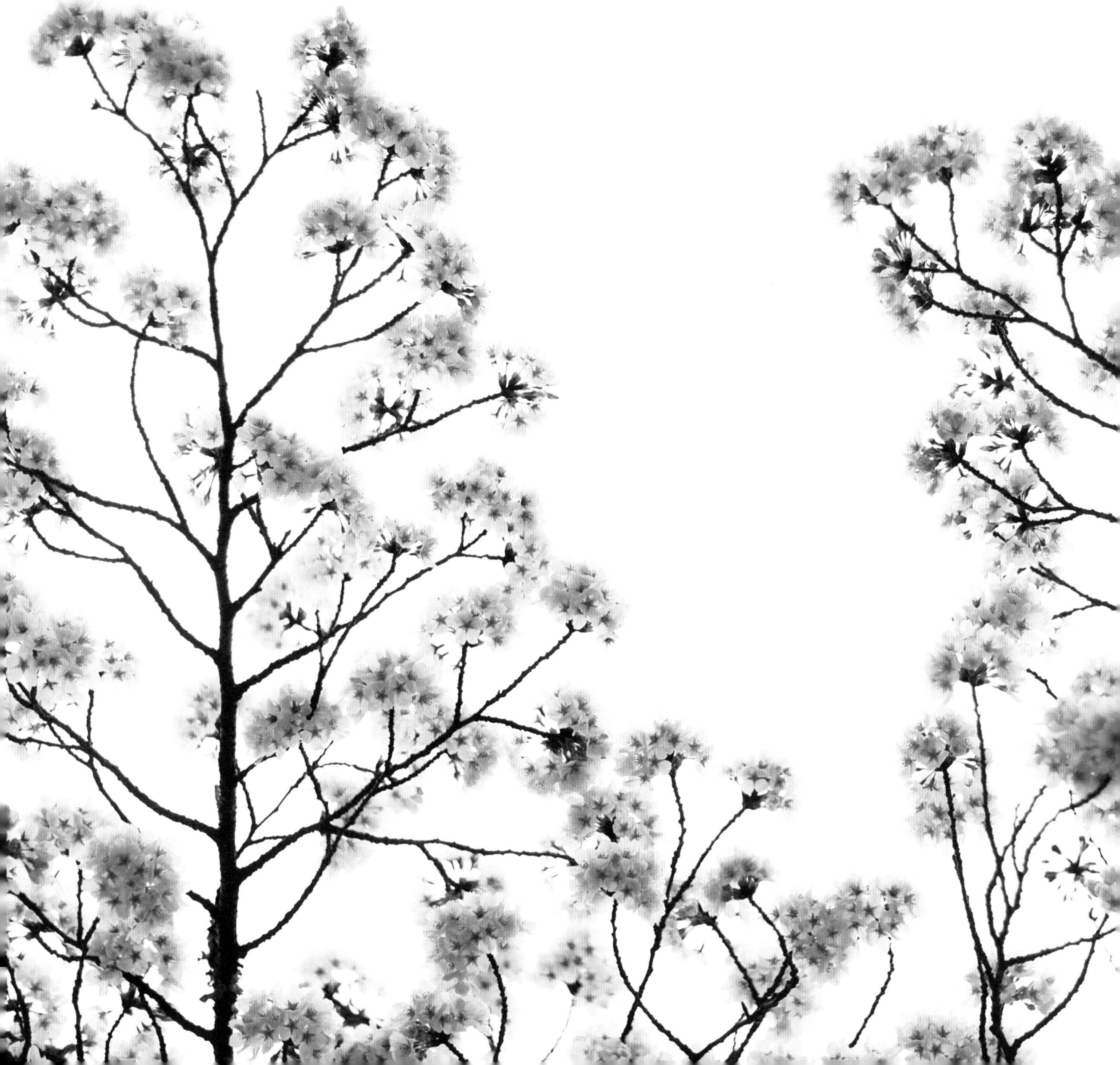

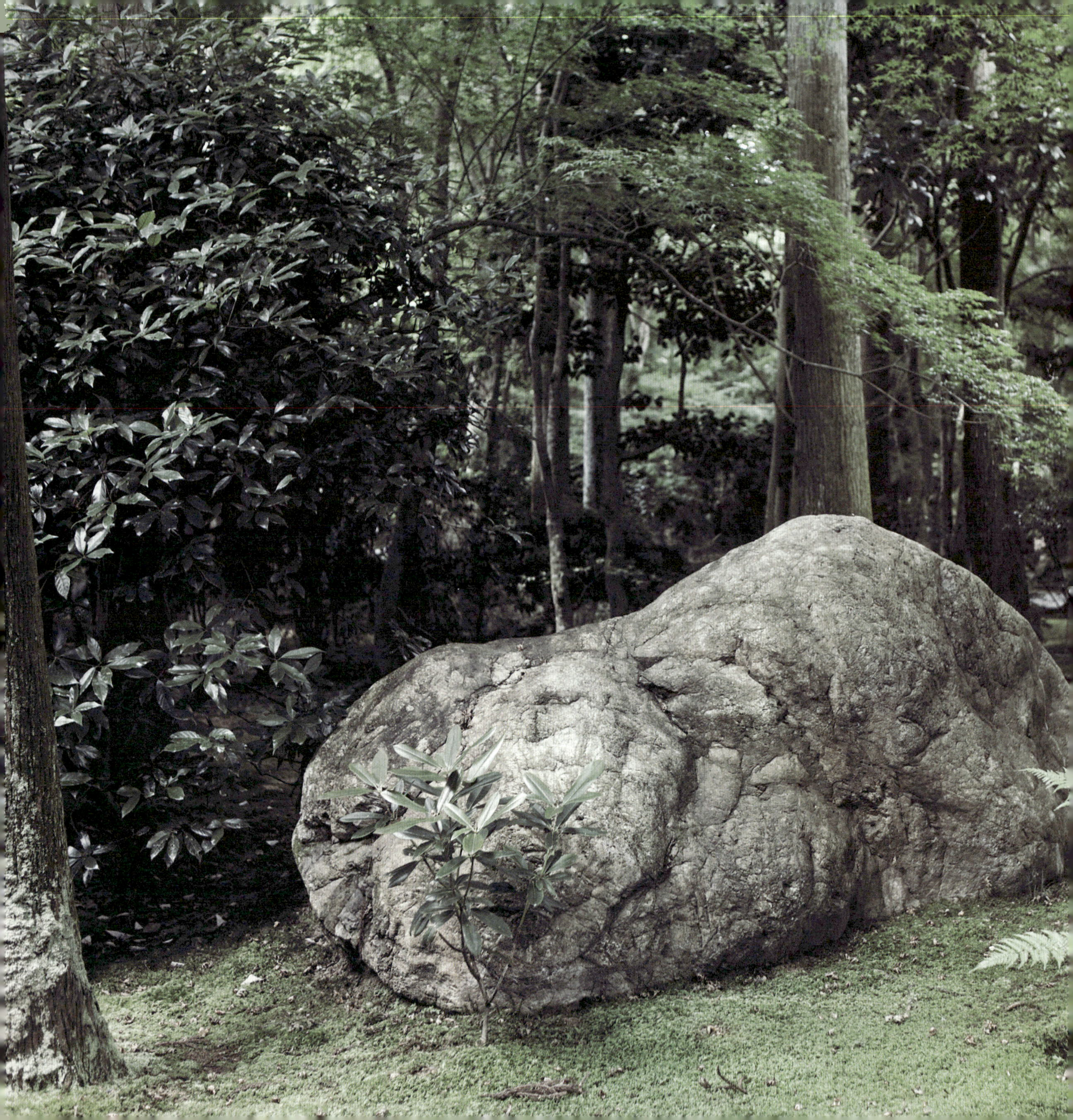

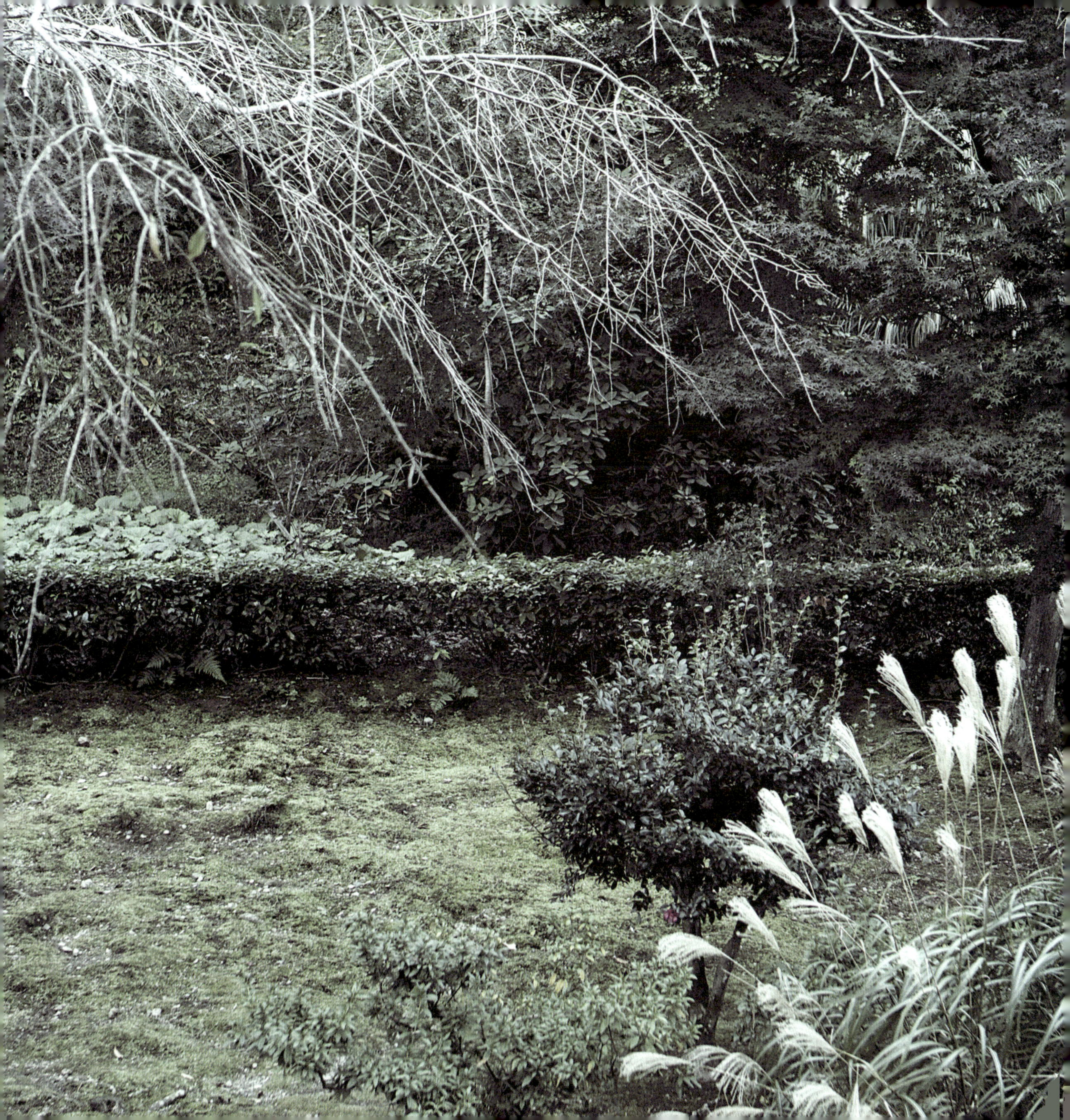

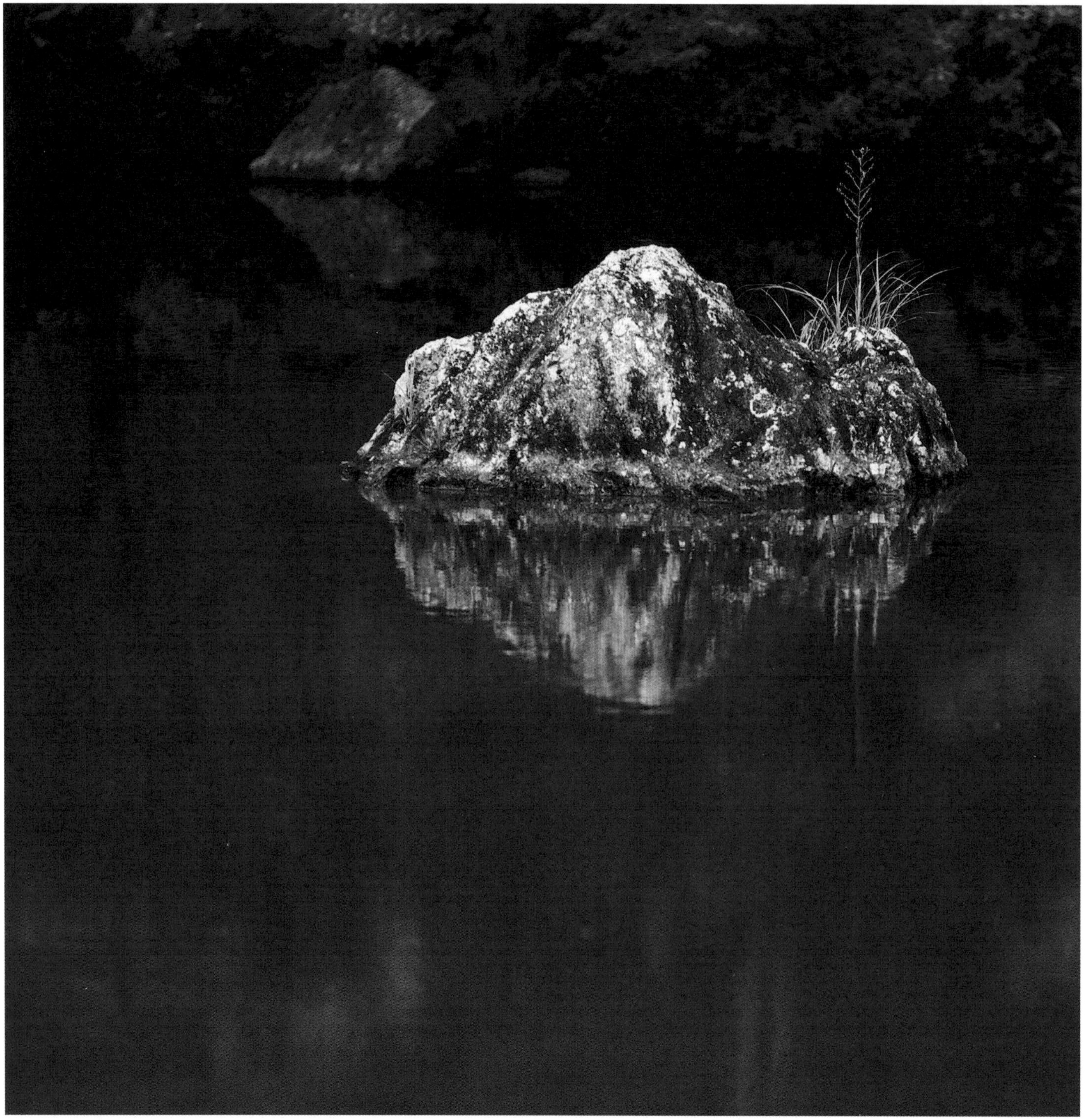

ANCIENT POND.
A FROG JUMPS IN. PLOPP.

Bashō Matsuo

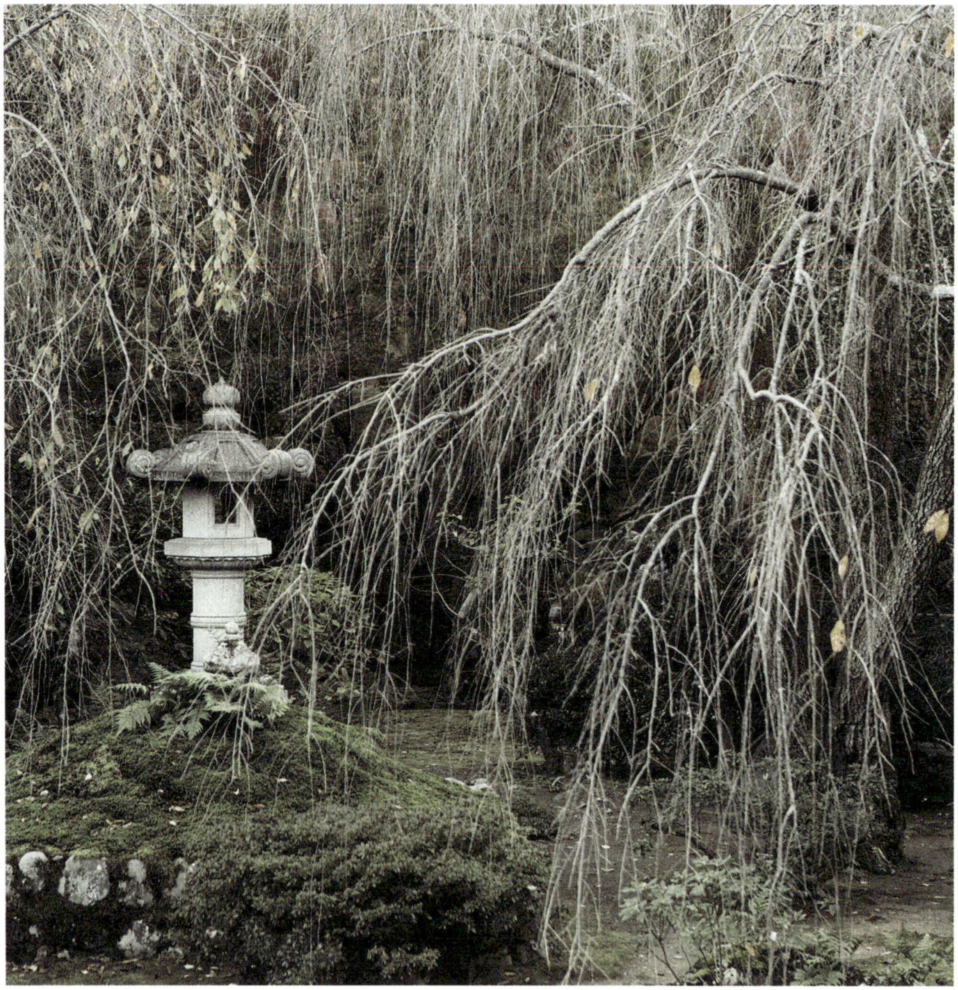

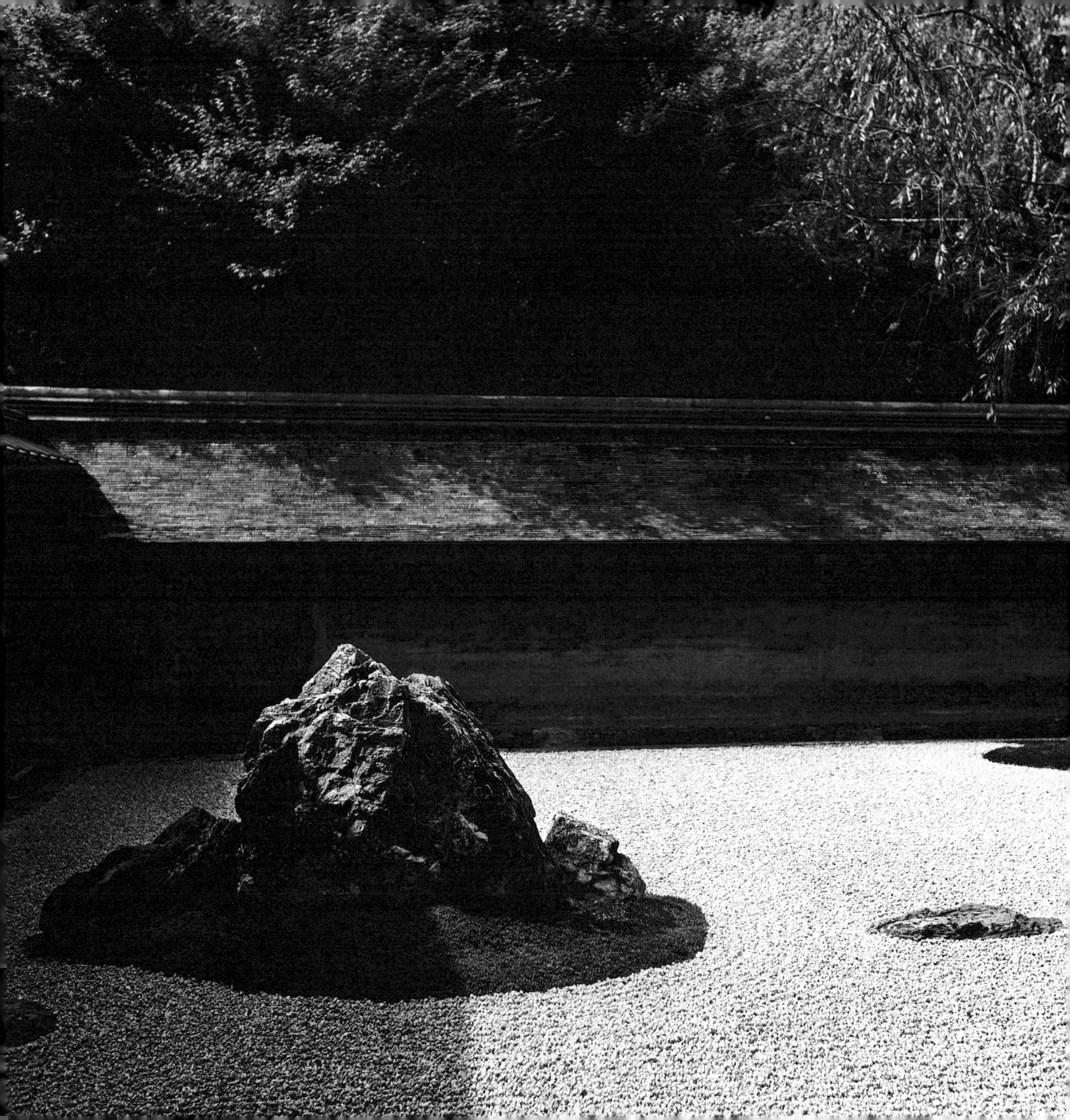

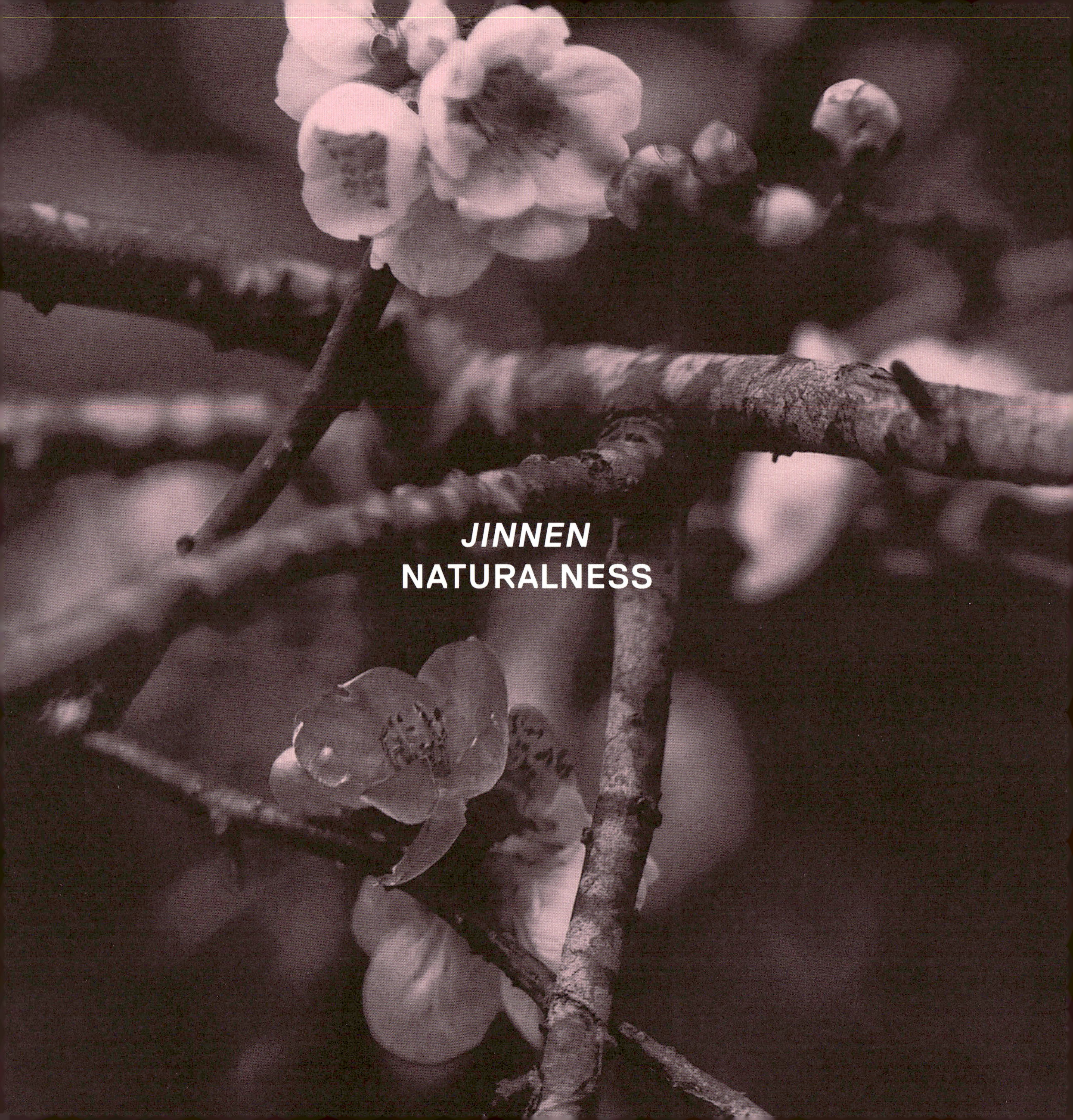

JINNEN
NATURALNESS

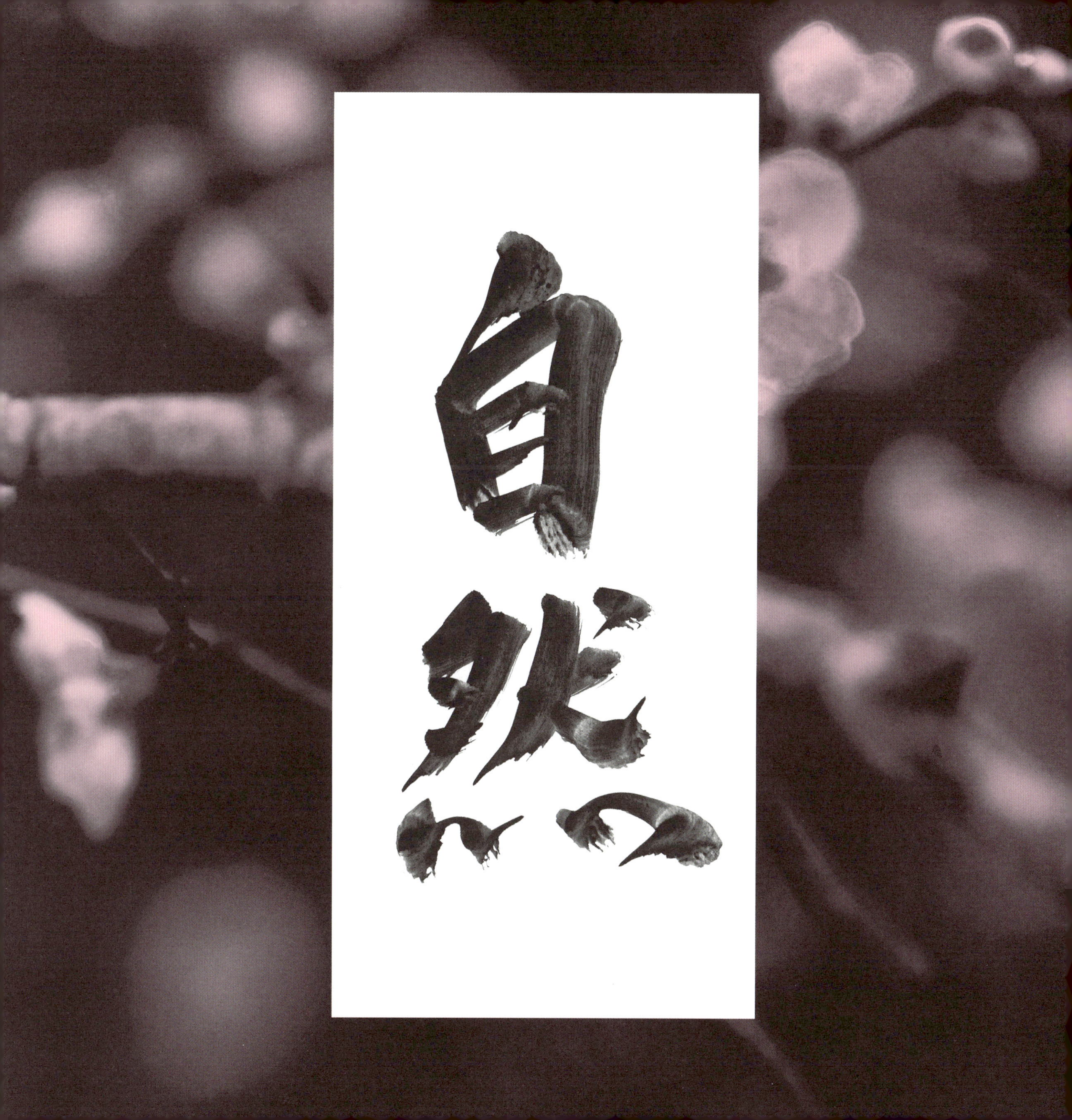
自然

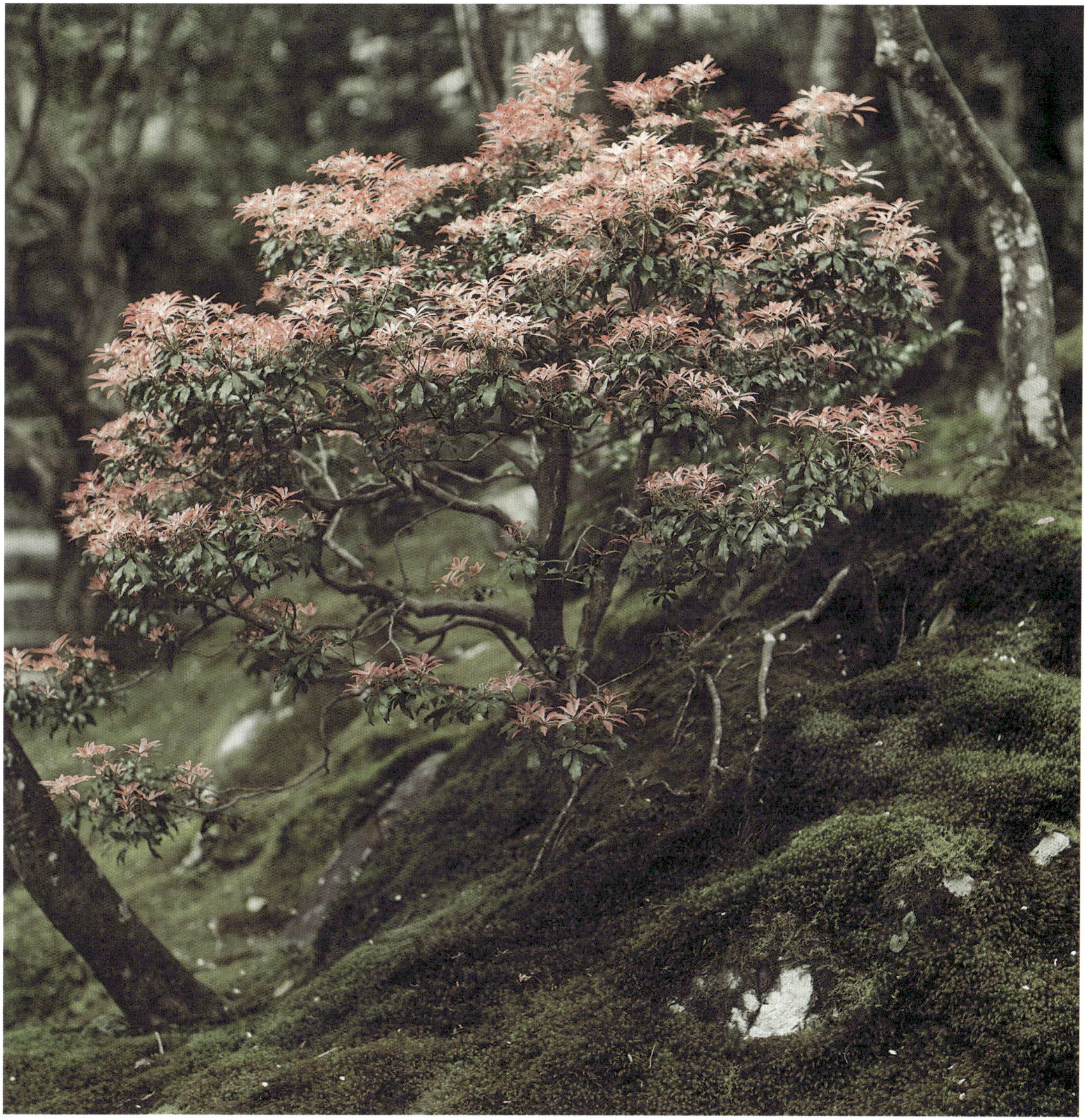

LANDSCAPE LOOKS APPETIZING.
THE SNOW SOFT AND GENTLE.

Kobayashi Issa

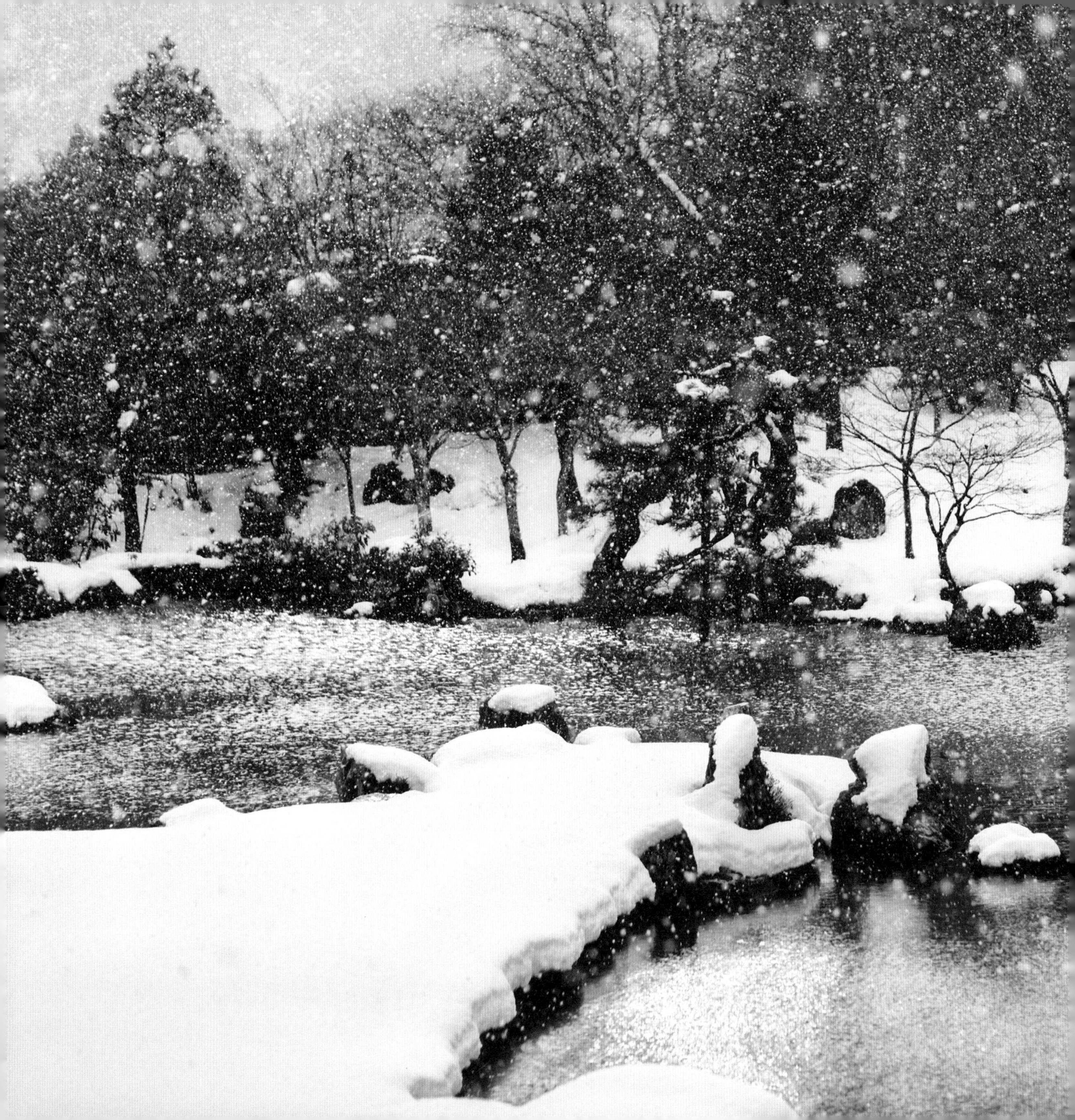

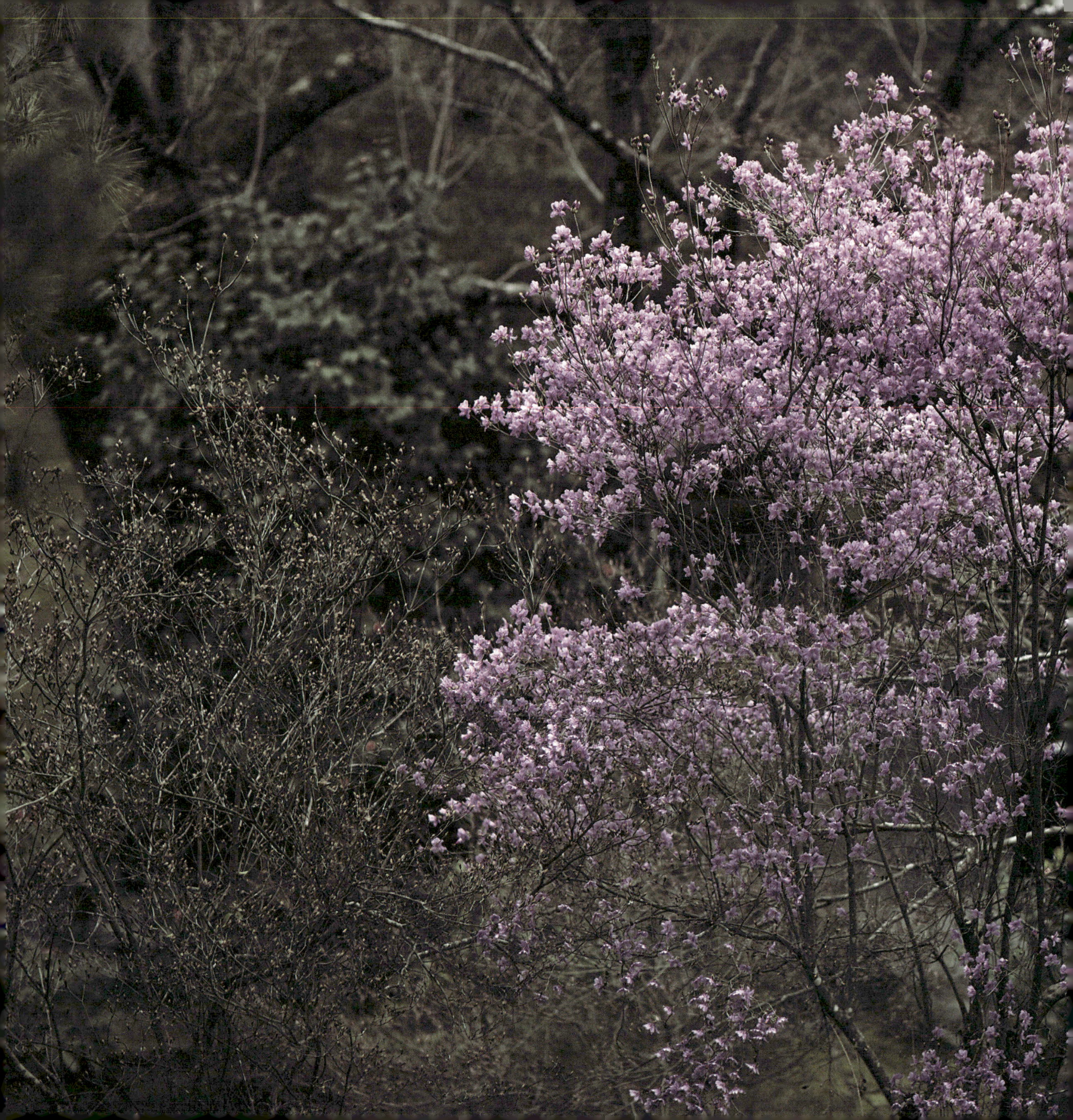

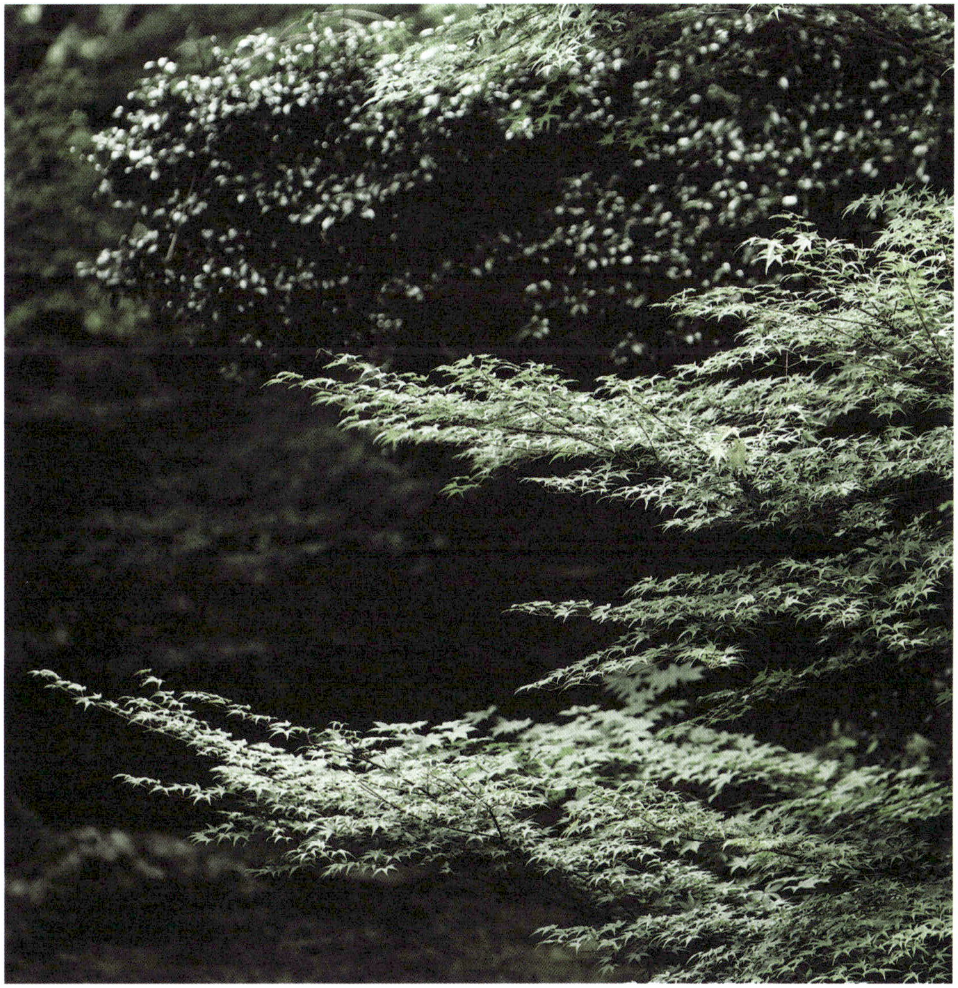

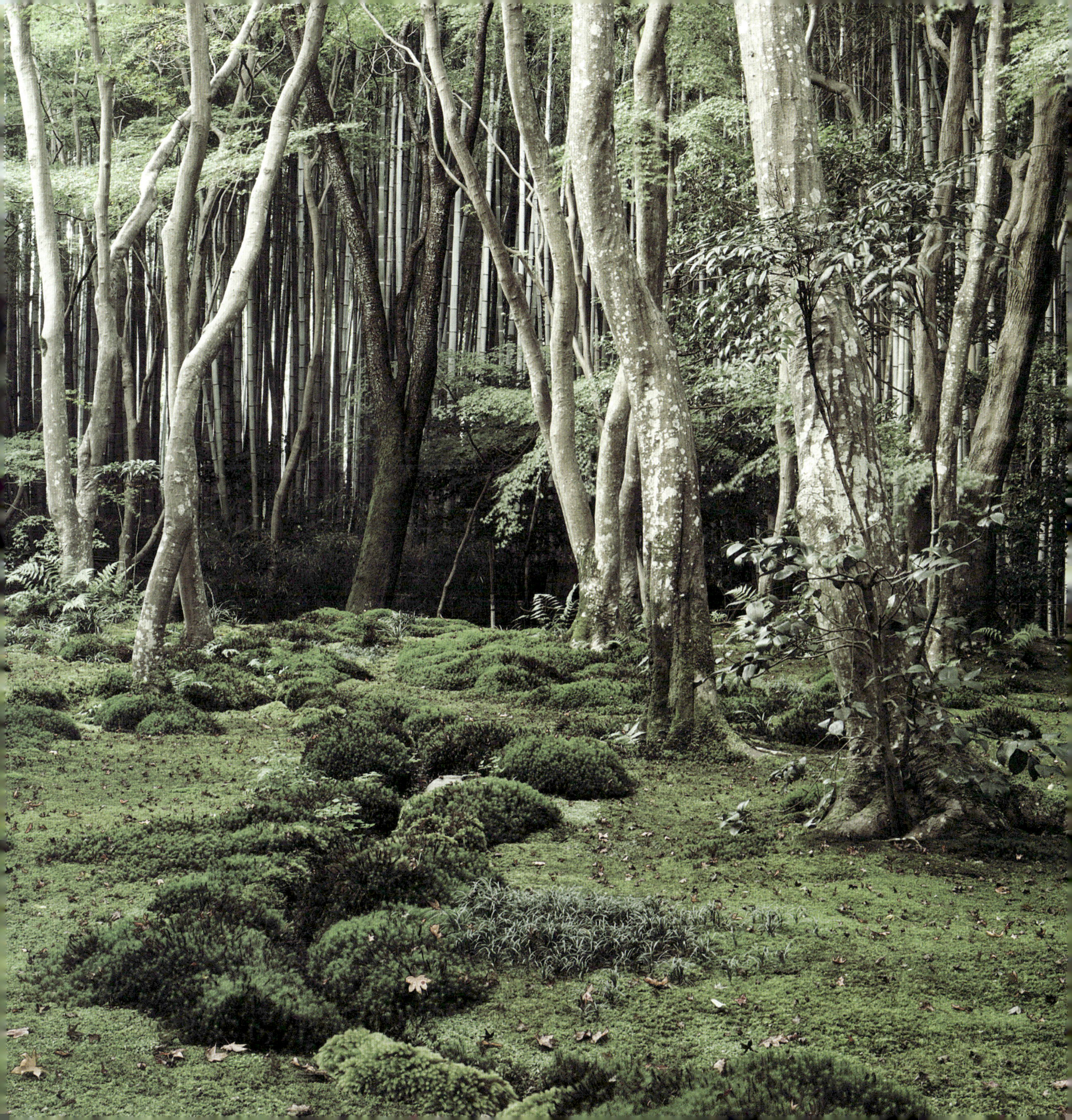

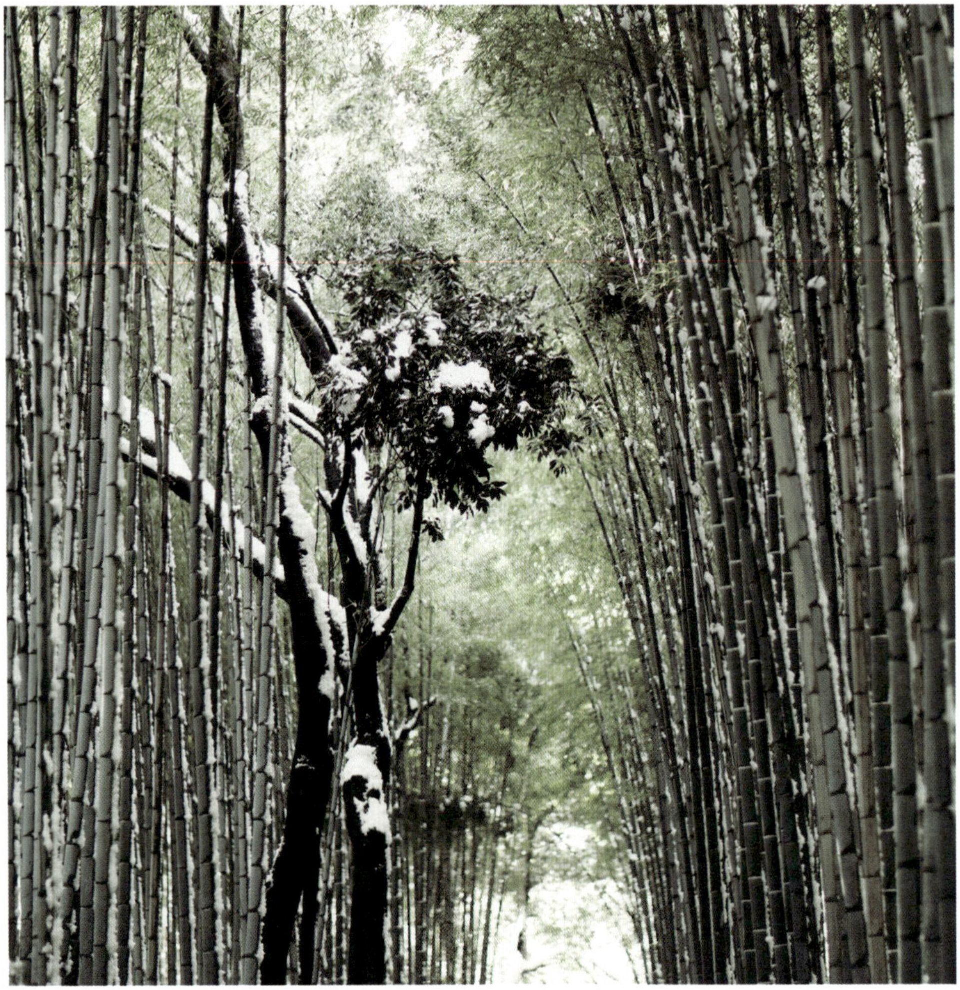

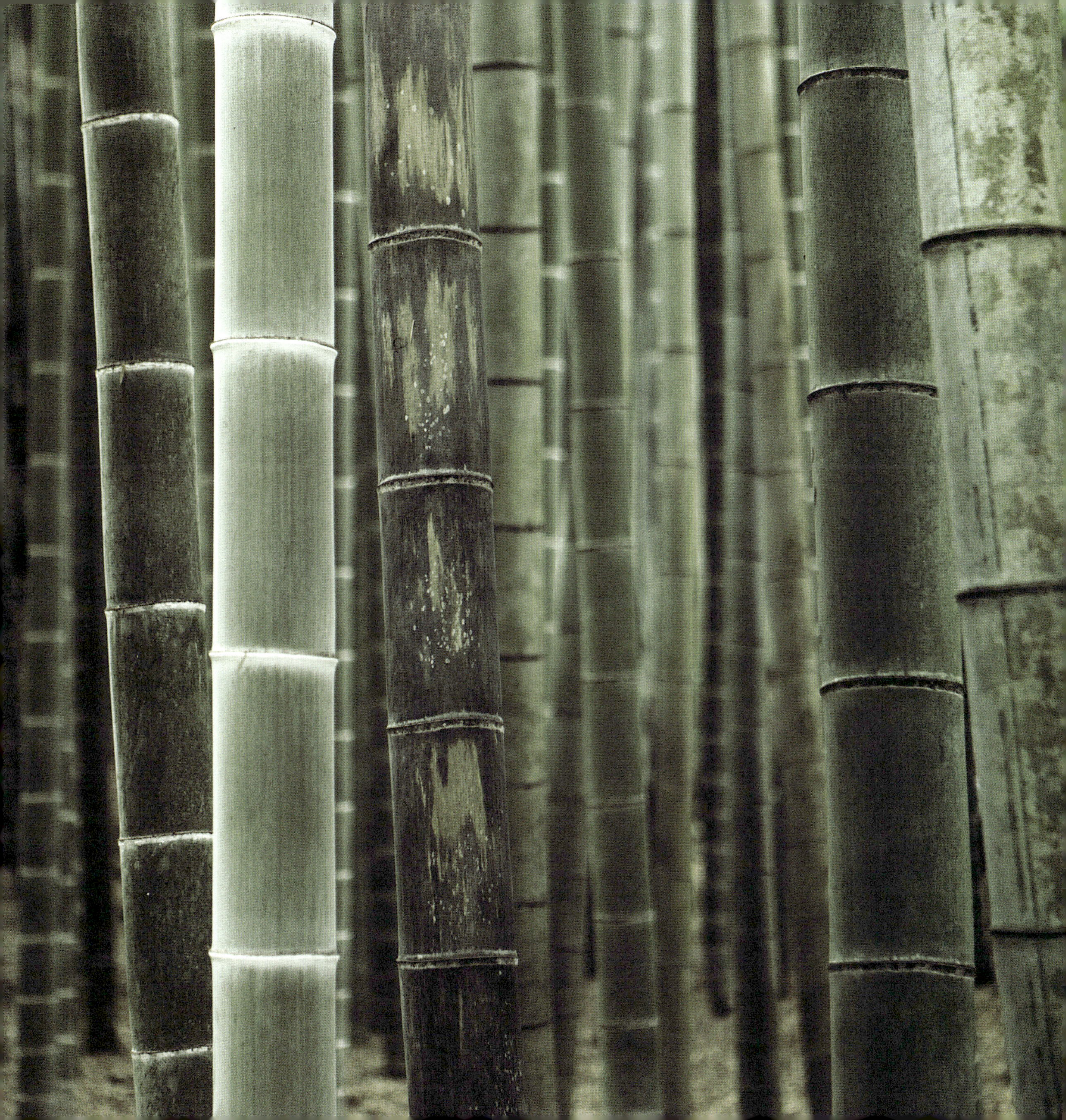

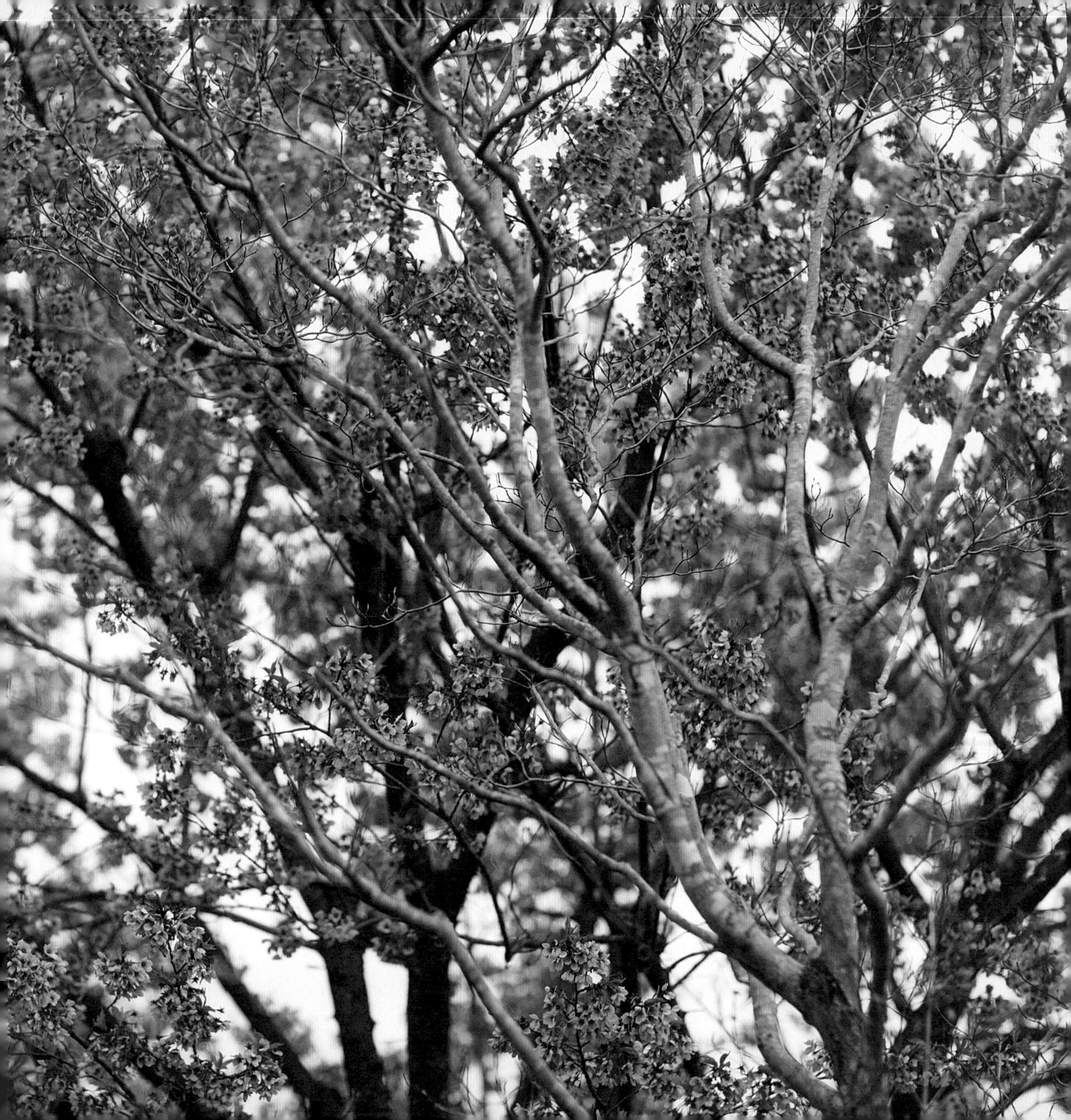

THERE IS NOTHING YOU CAN SEE
THAT IS NOT A FLOWER.
THERE IS NOTHING YOU CAN THINK
THAT IS NOT THE MOON.

Bashō Matsuo

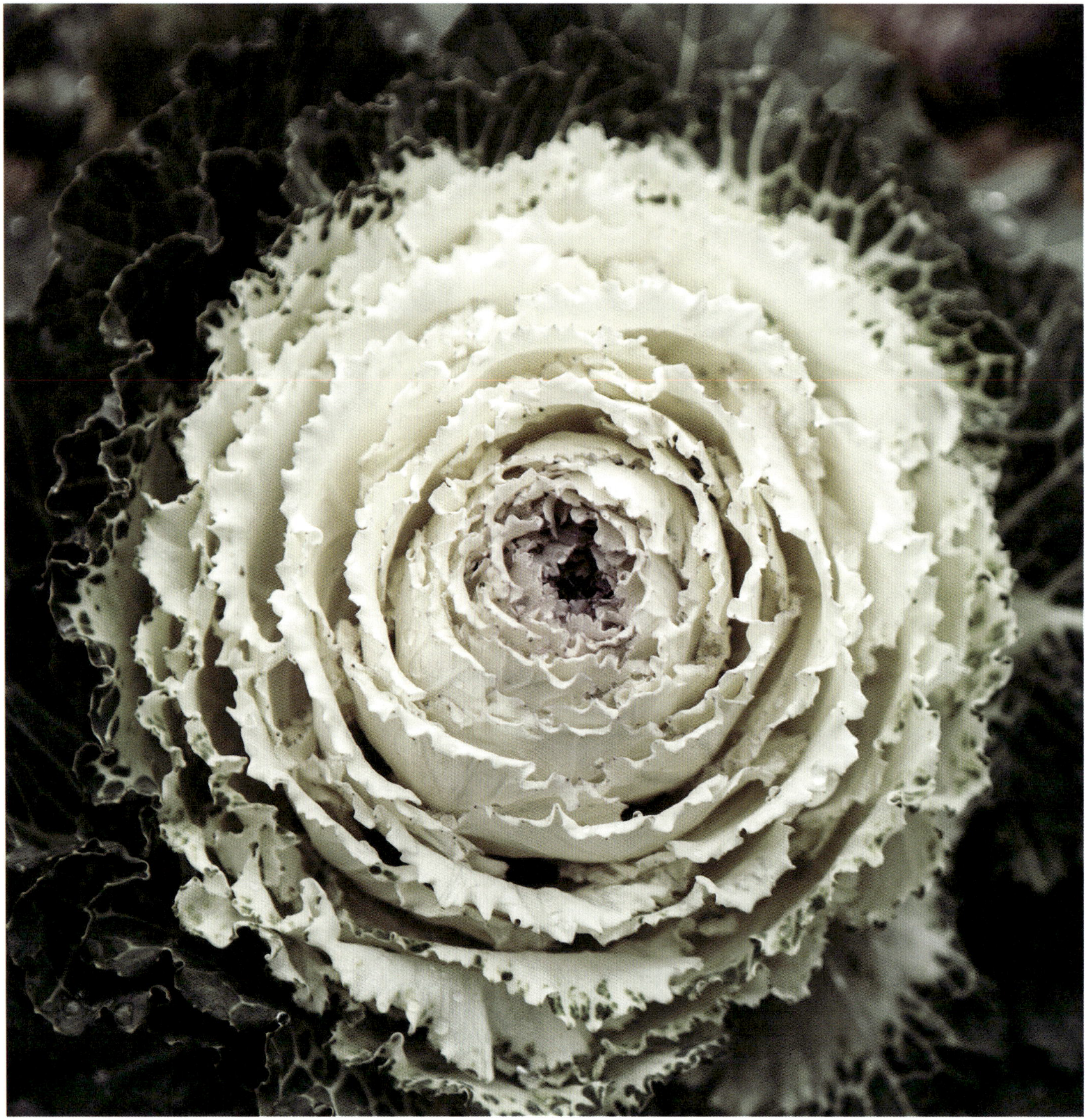

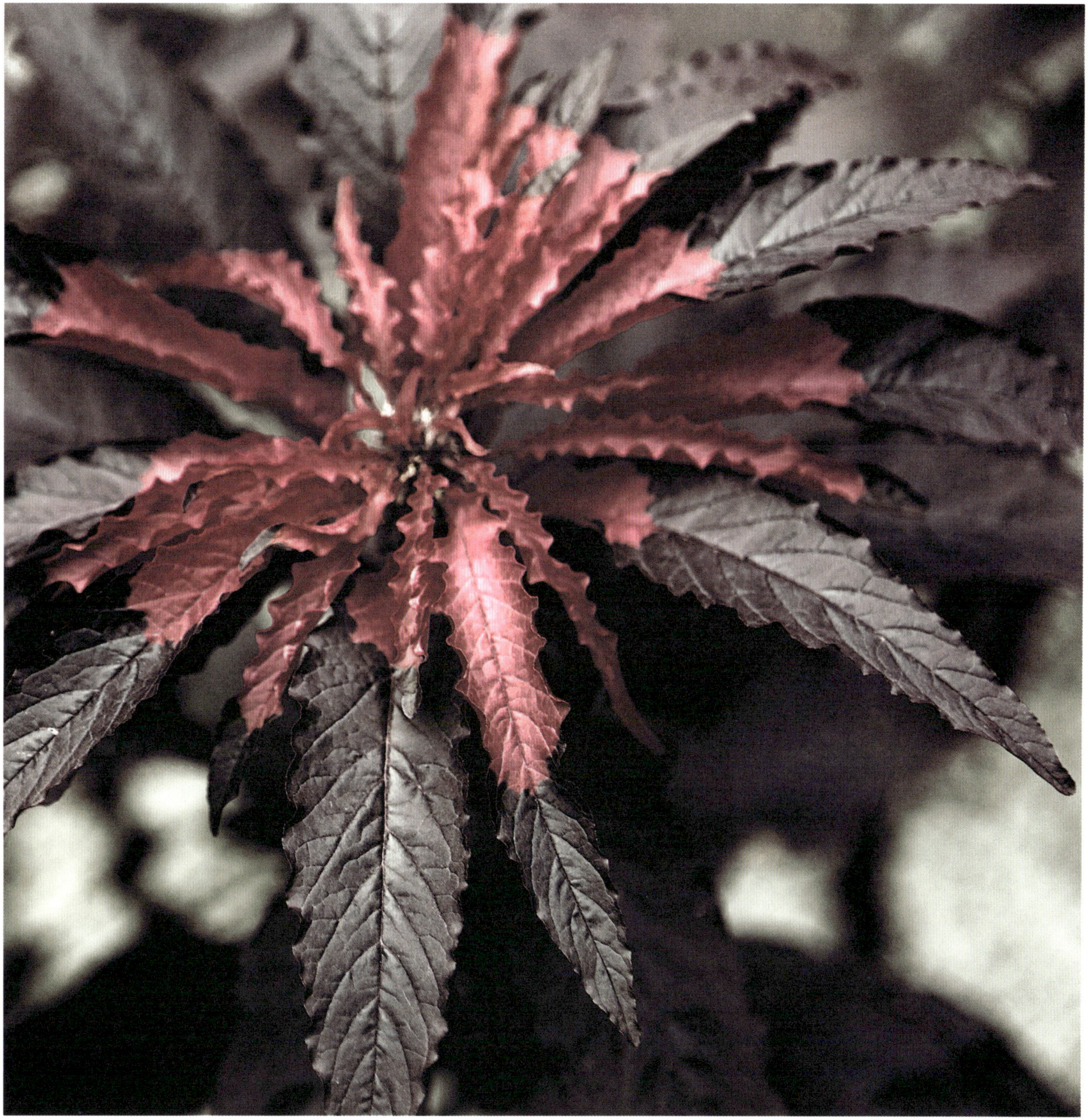

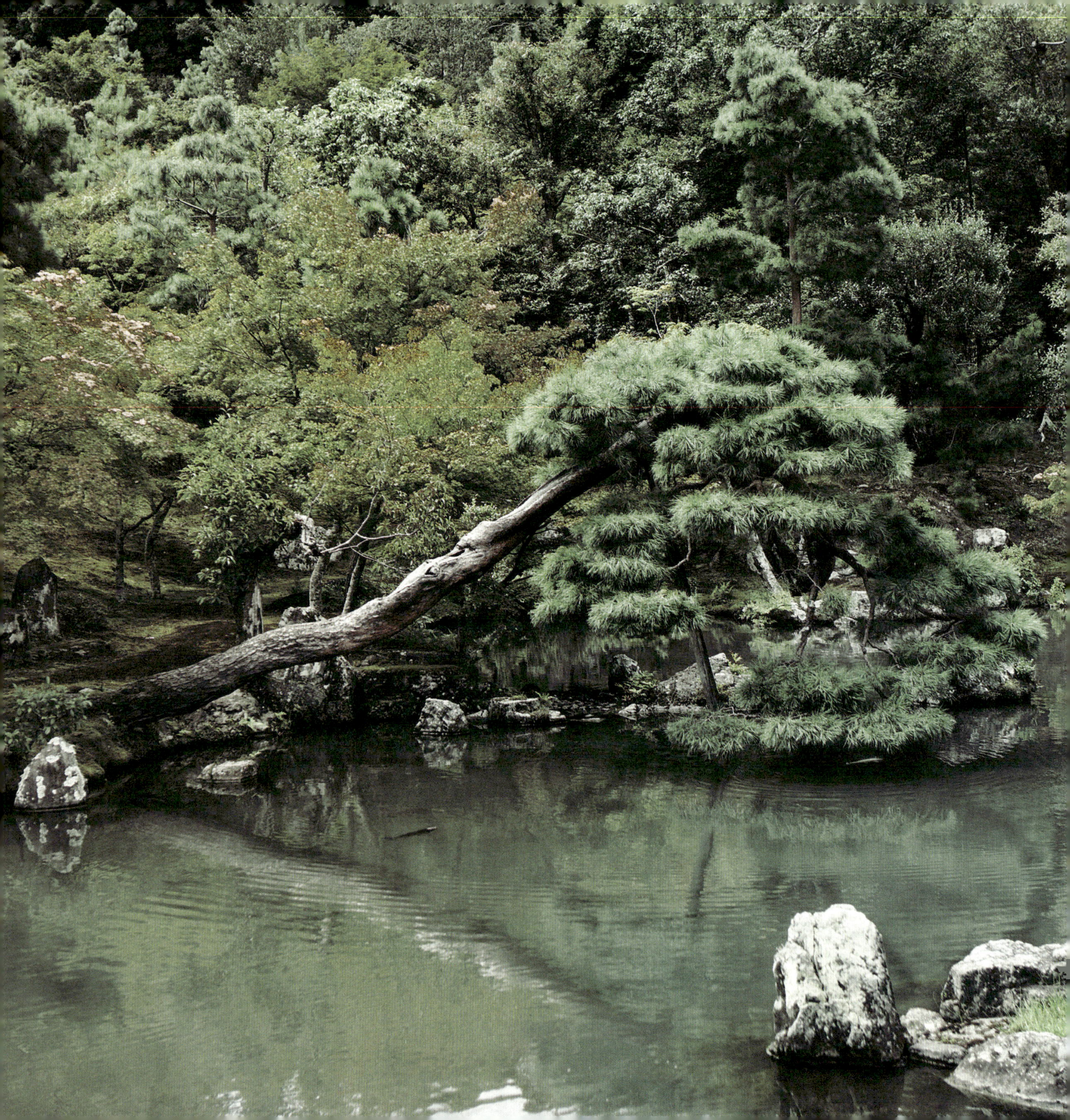

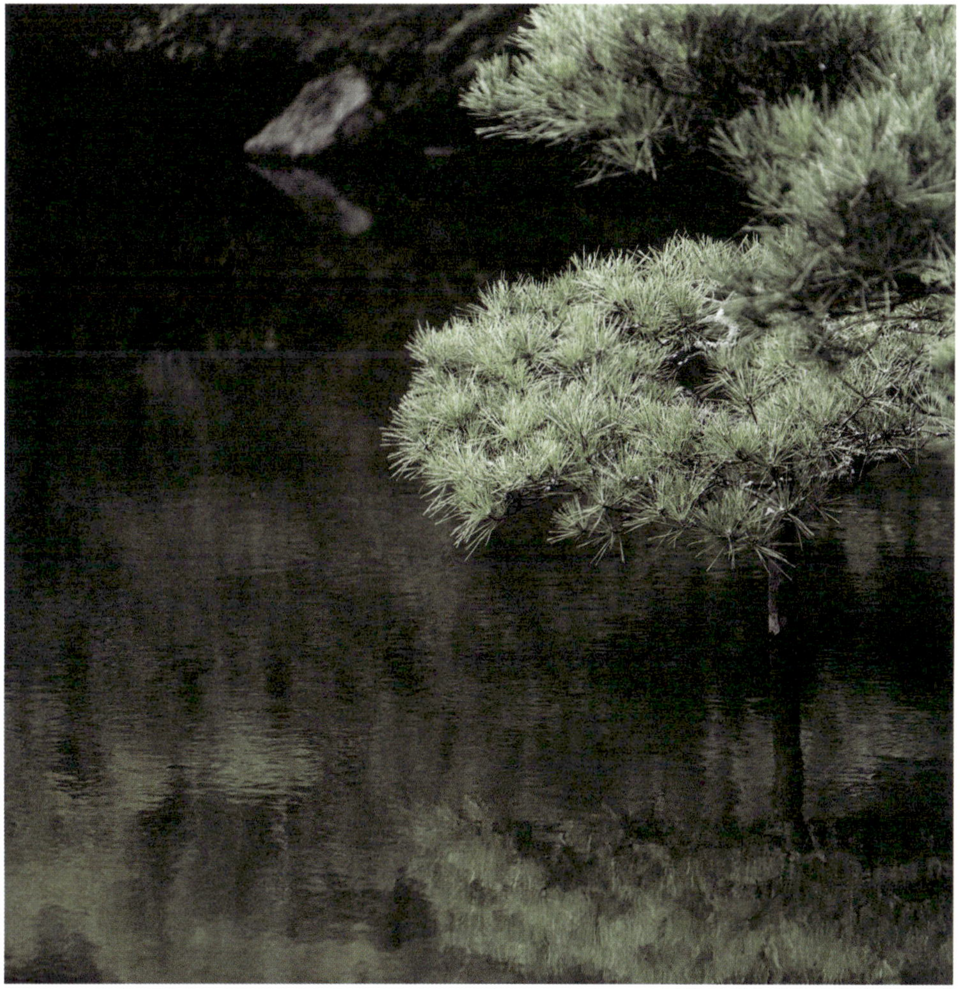

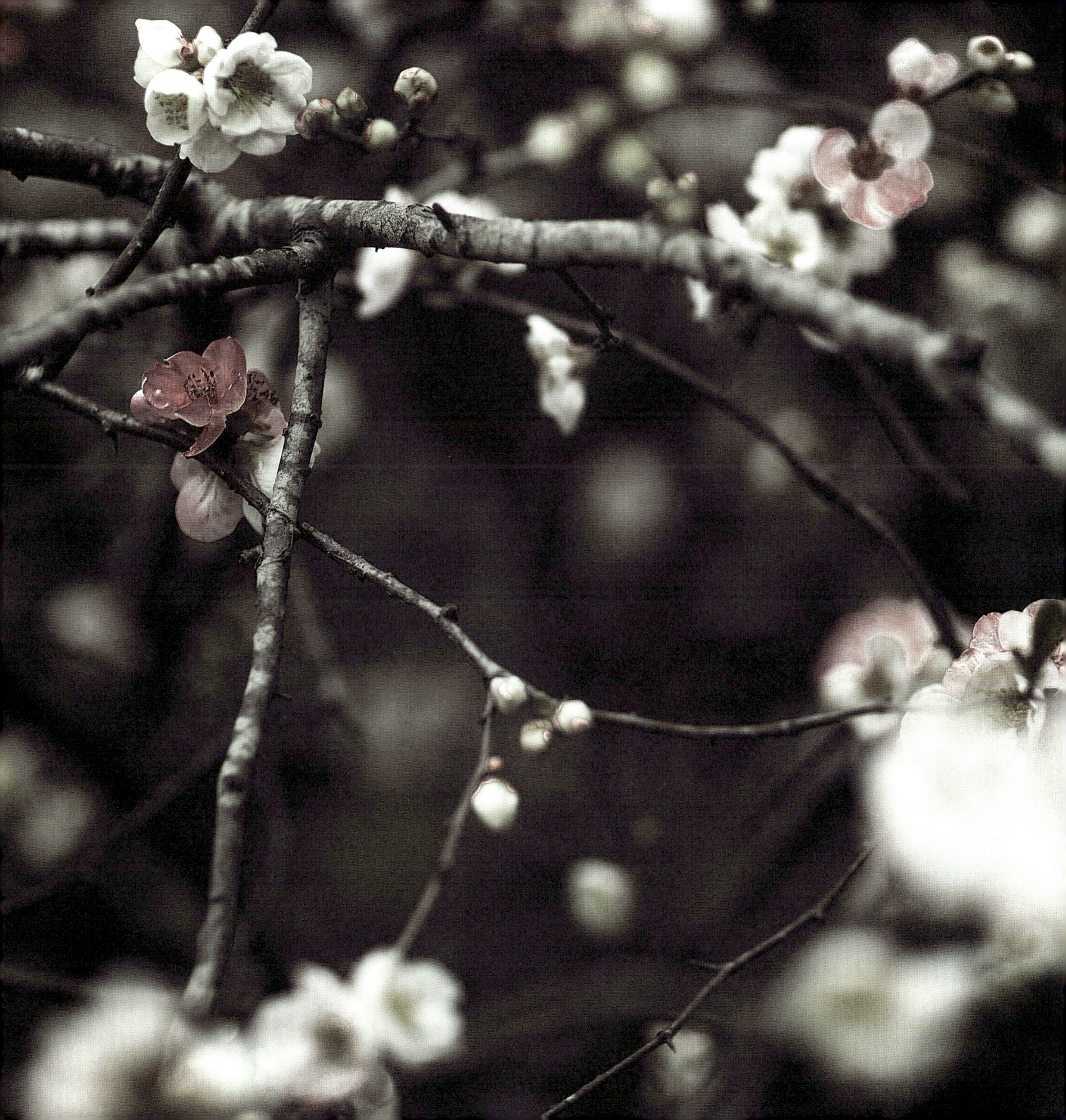

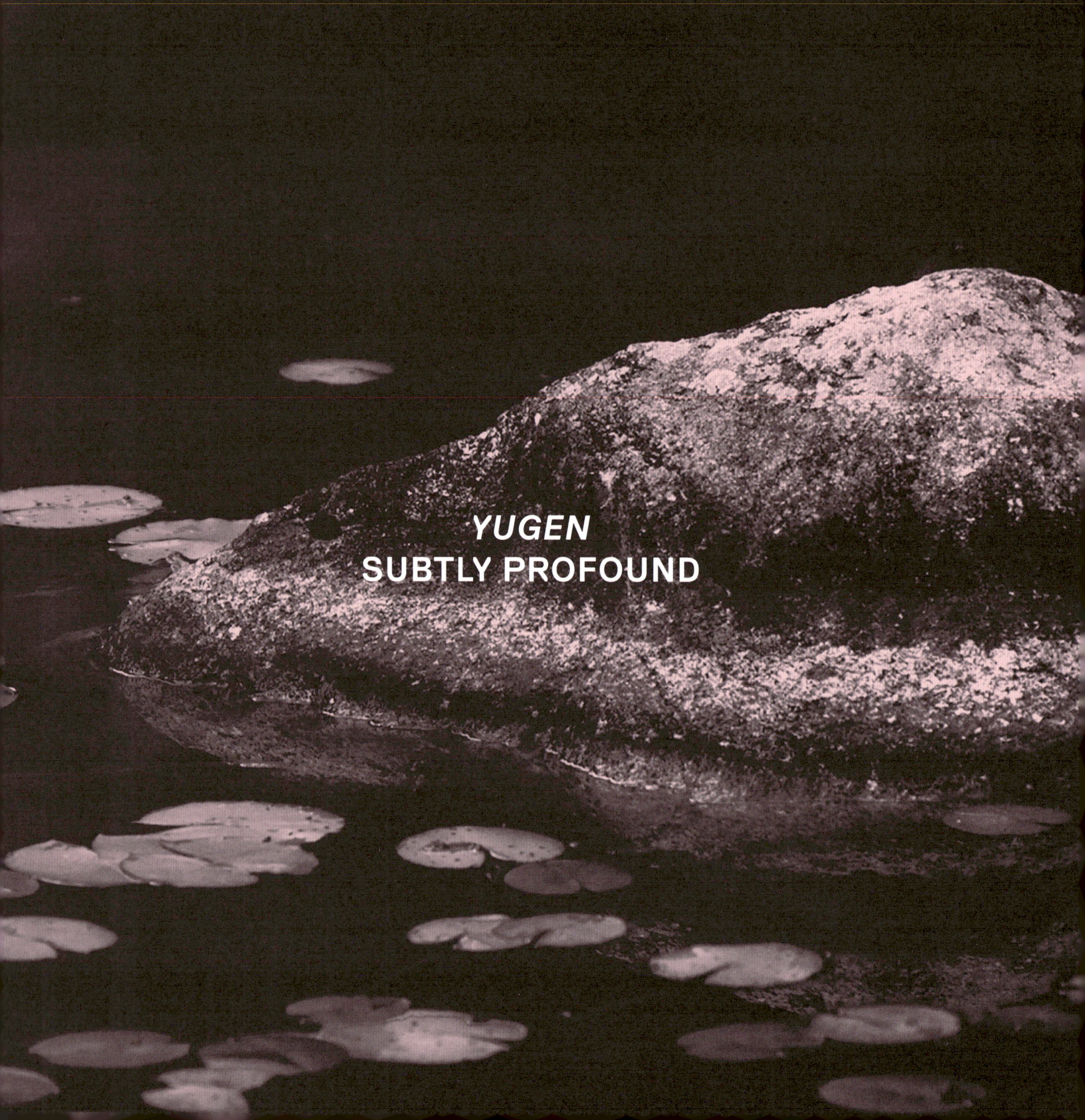

YUGEN
SUBTLY PROFOUND

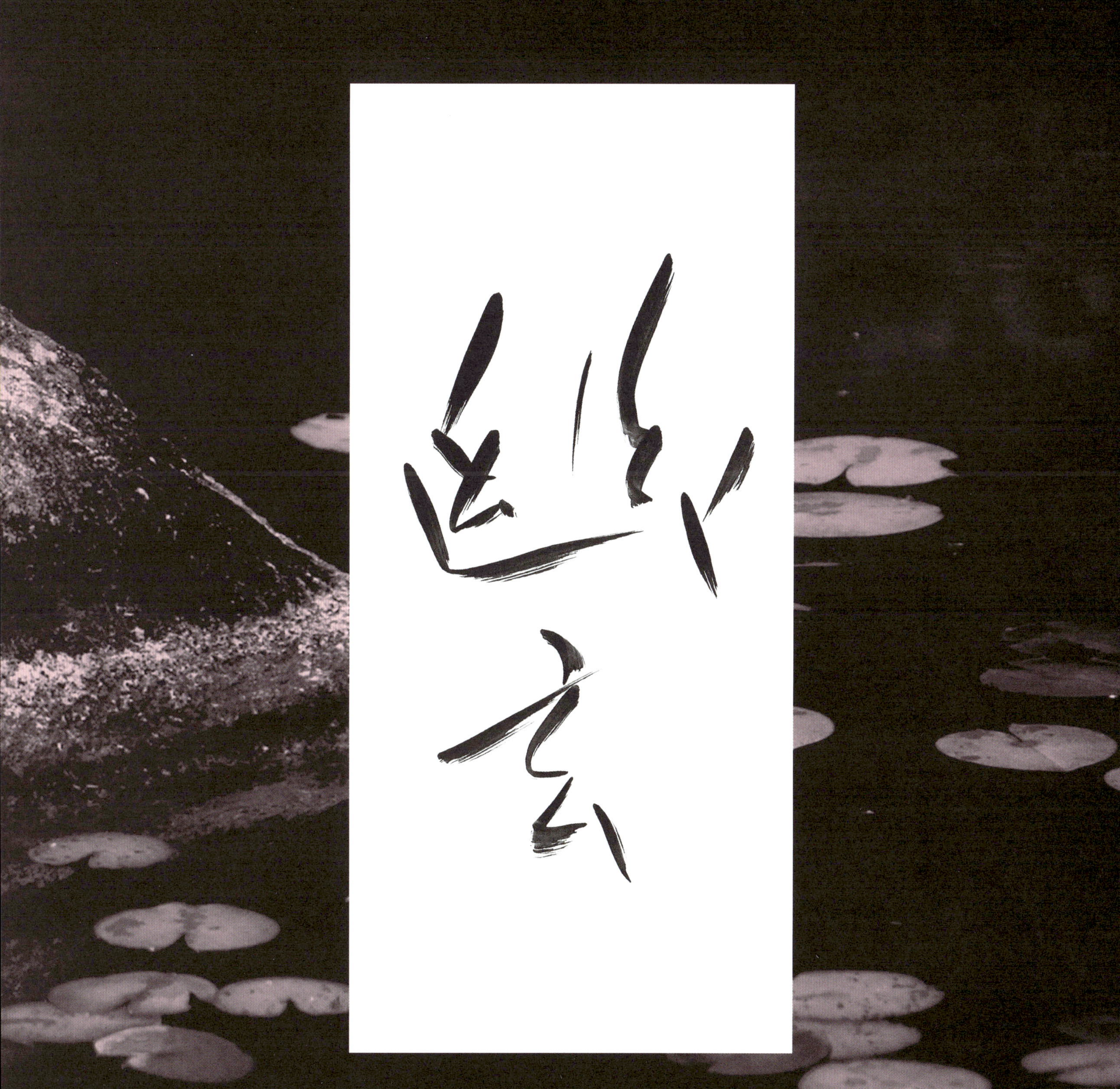

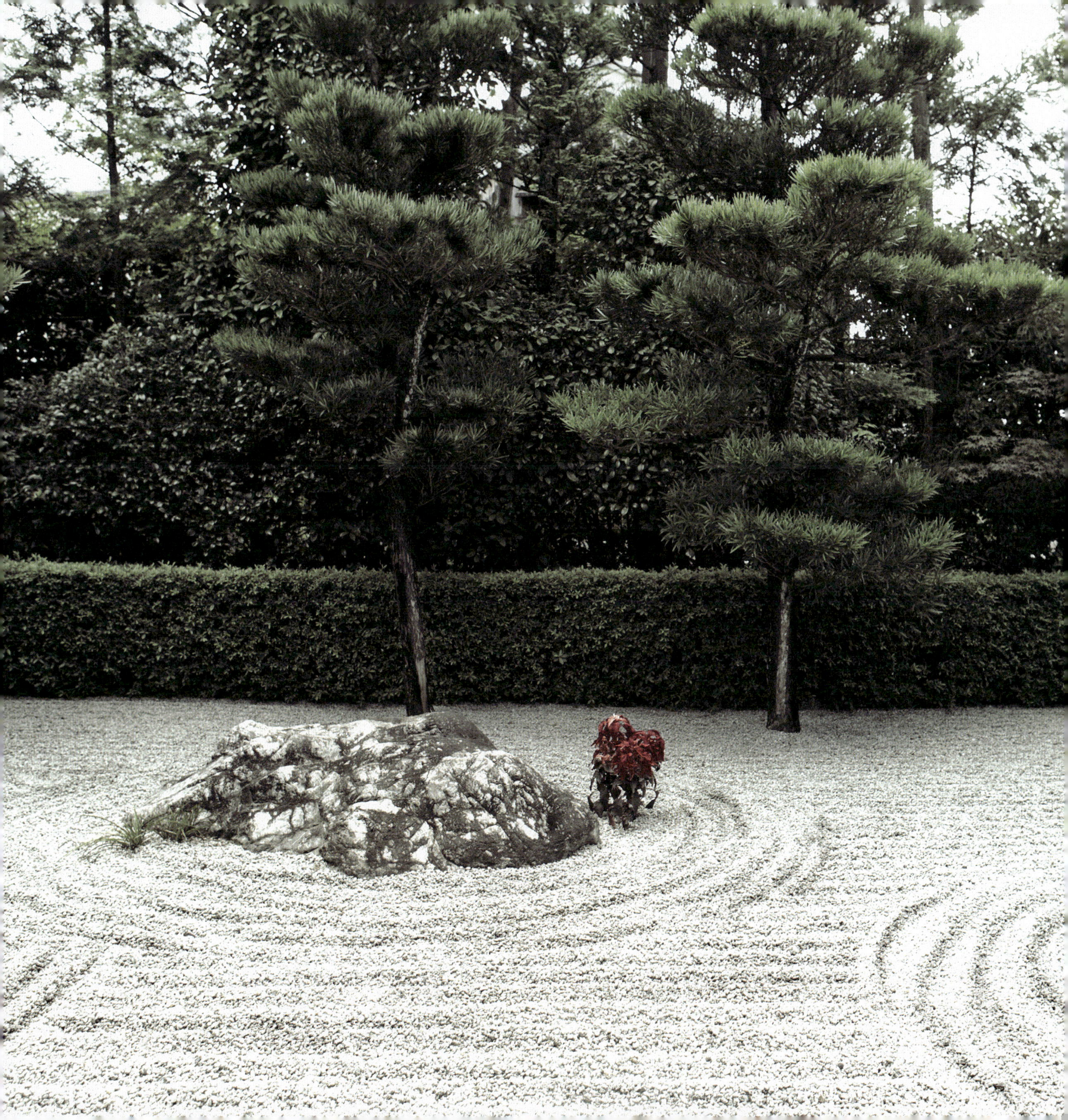

THE SHORT SUMMER NIGHT.
THE DREAM AND REAL ARE SAME THINGS.

Takahama Kyoshi

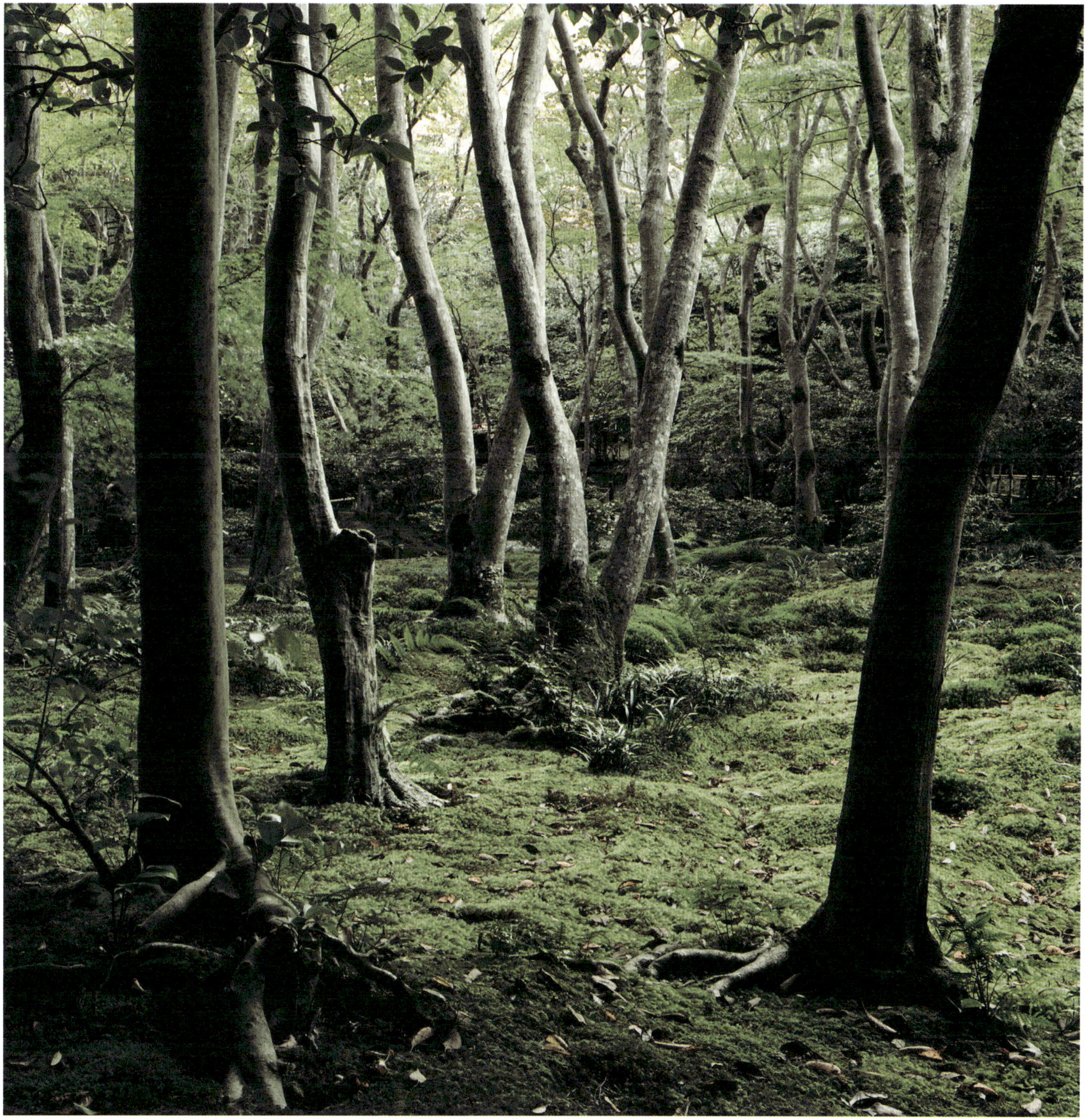

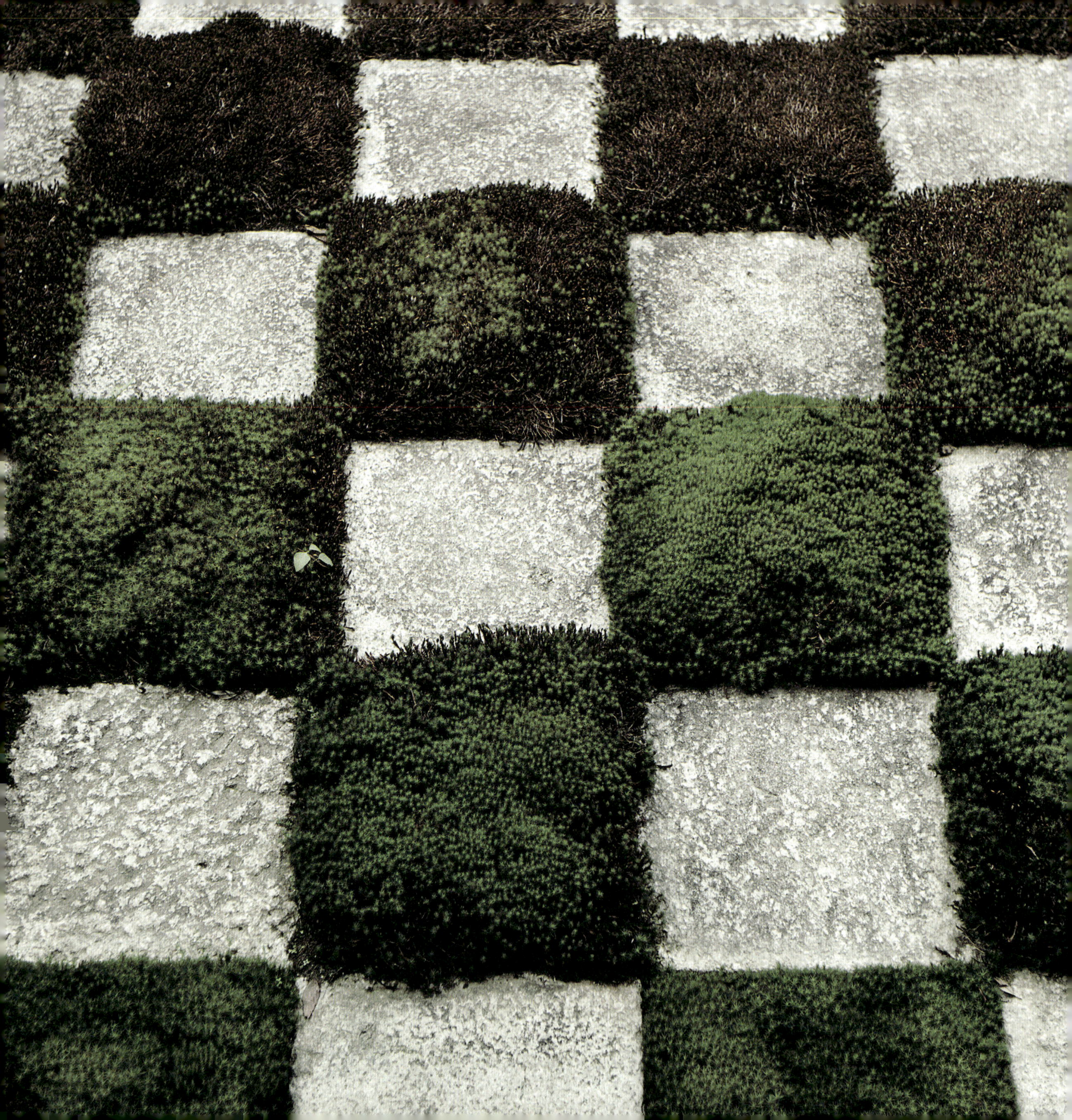

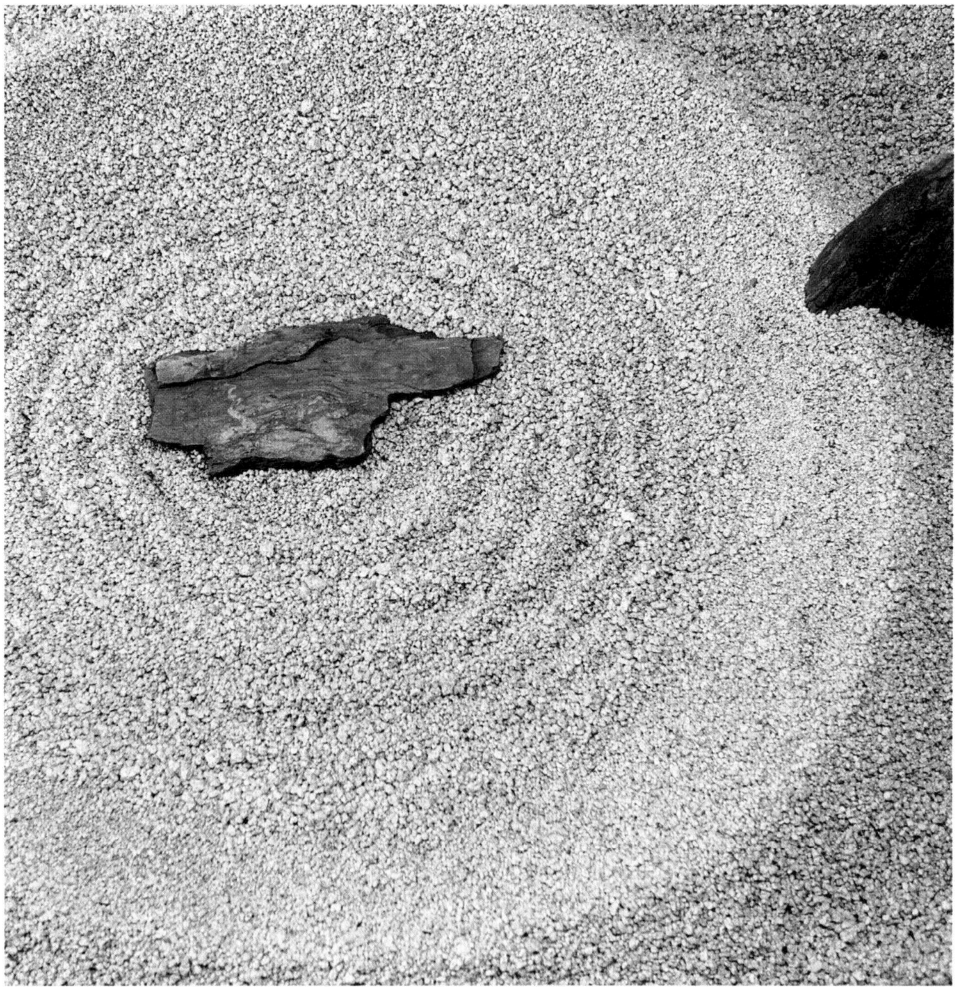

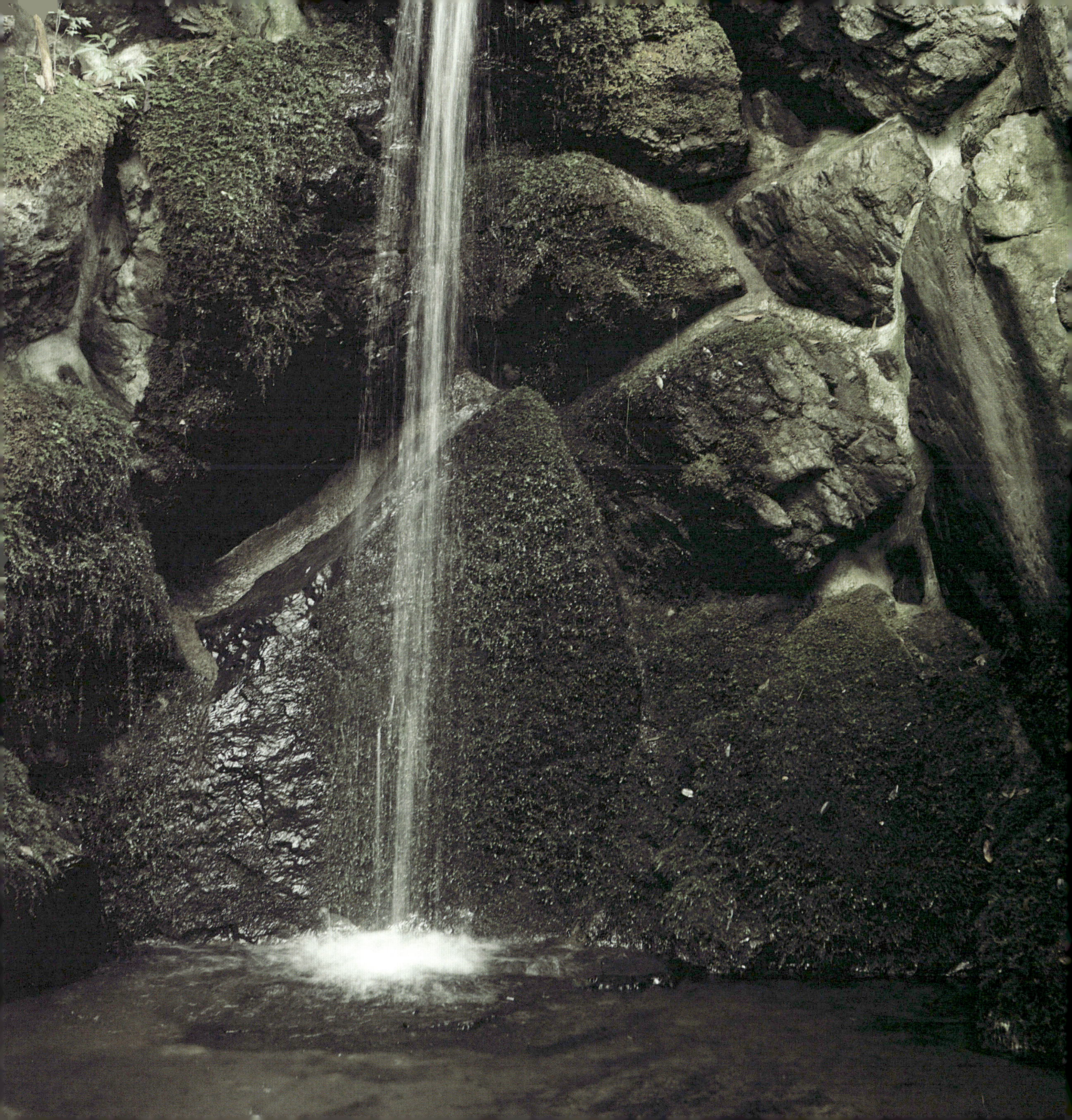

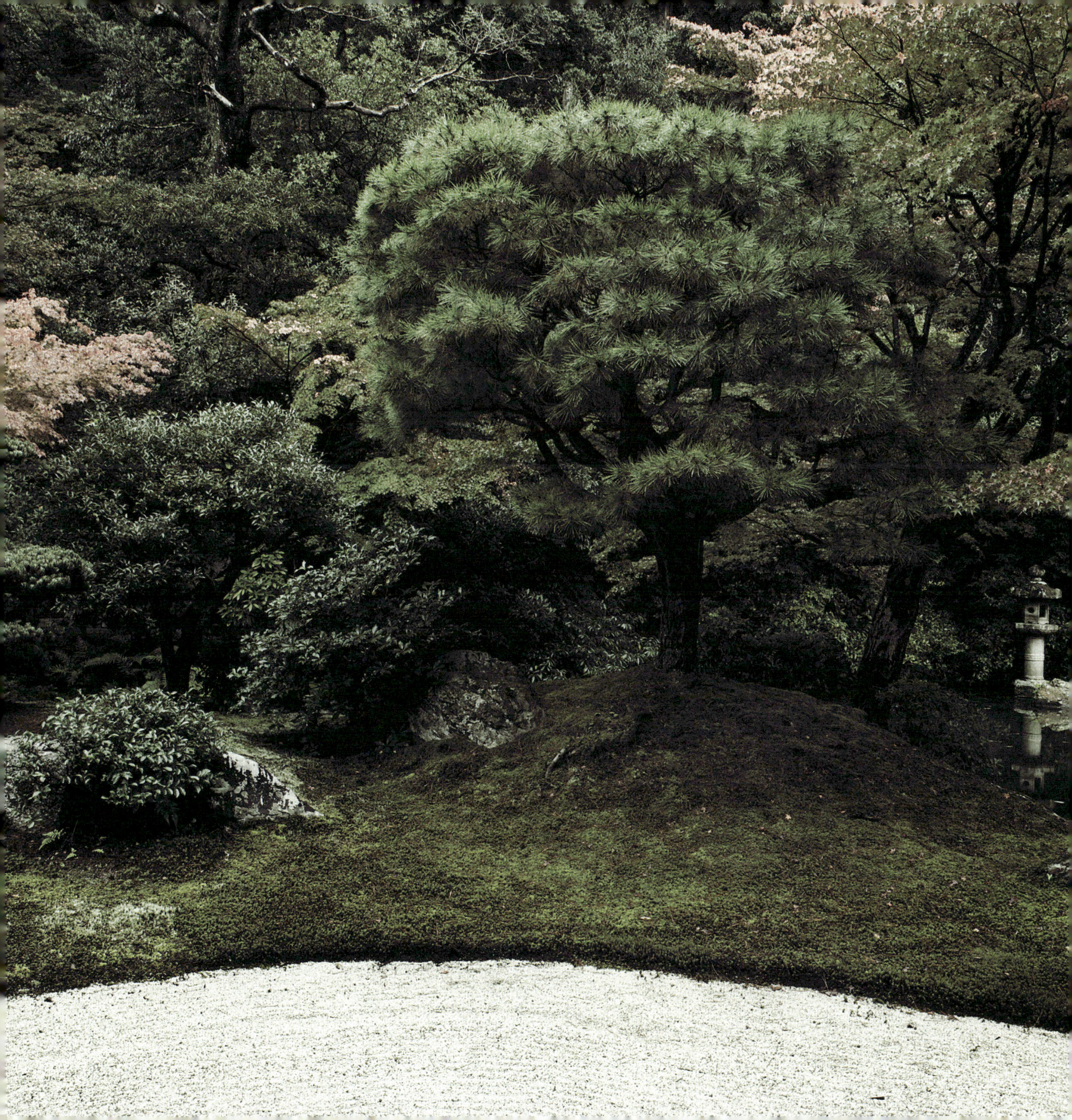

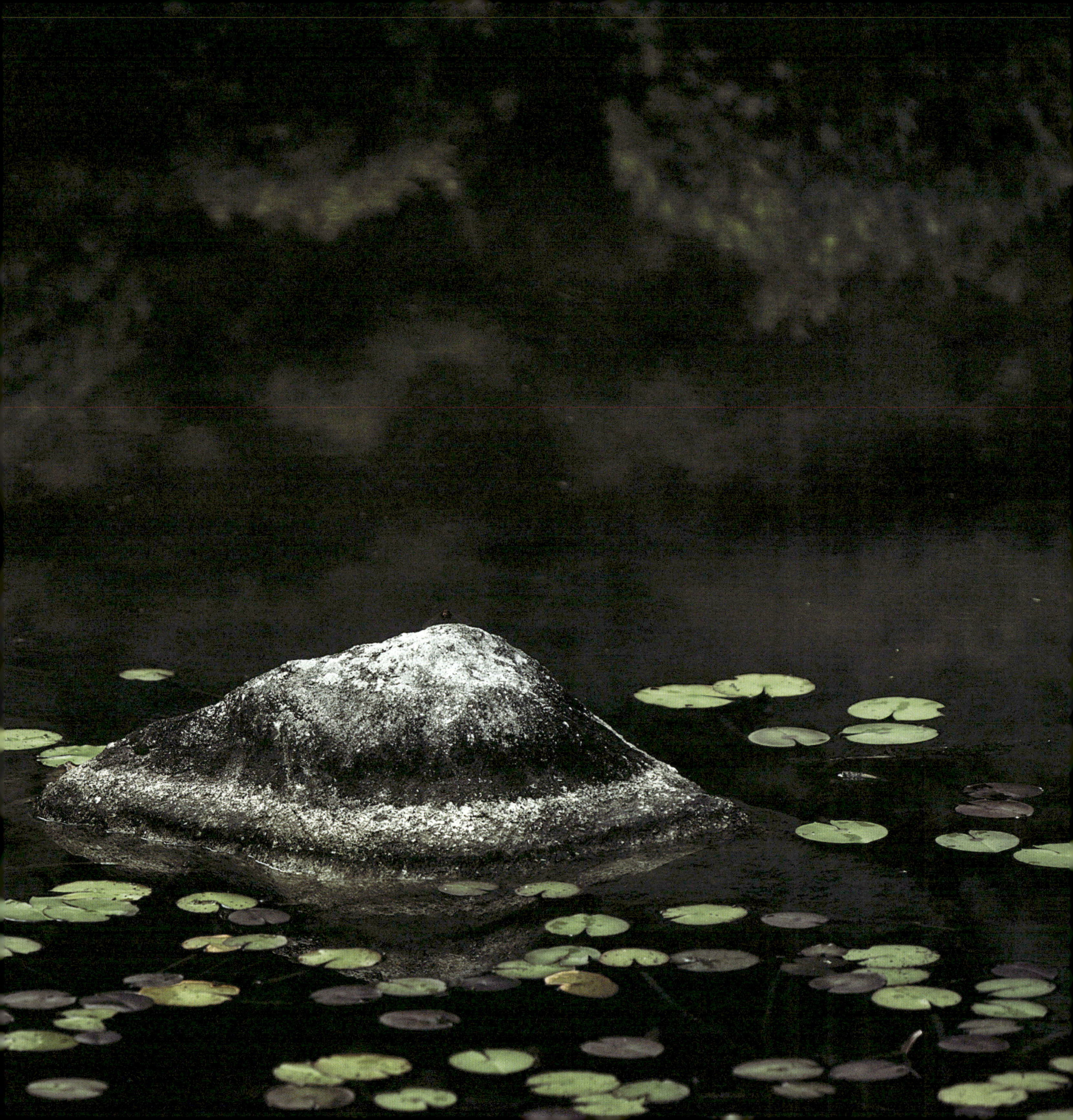

IN KYOTO HEARING THE
CUCKOO'S CRY I LONG FOR KYOTO.

Bashō Matsuo

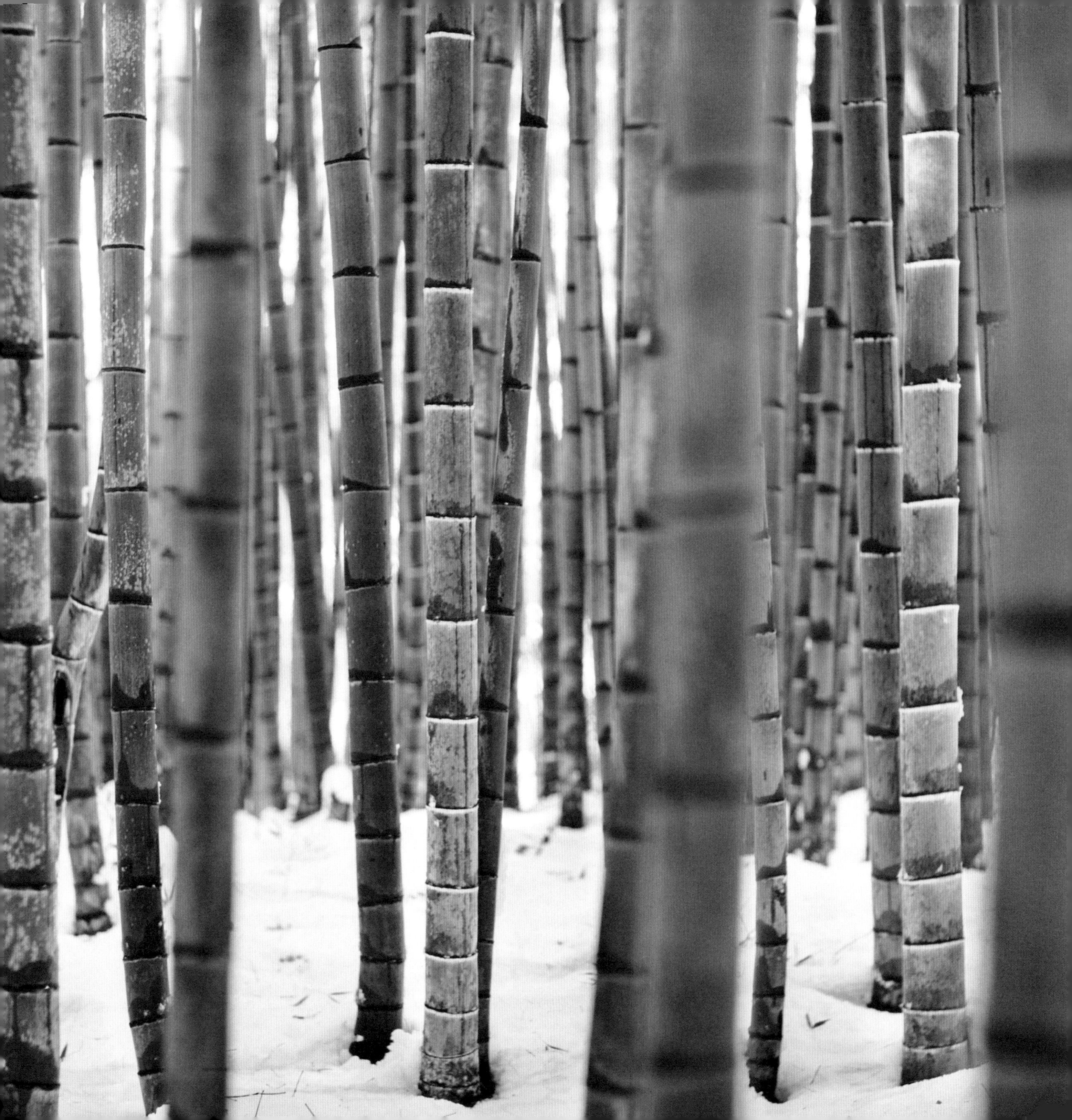

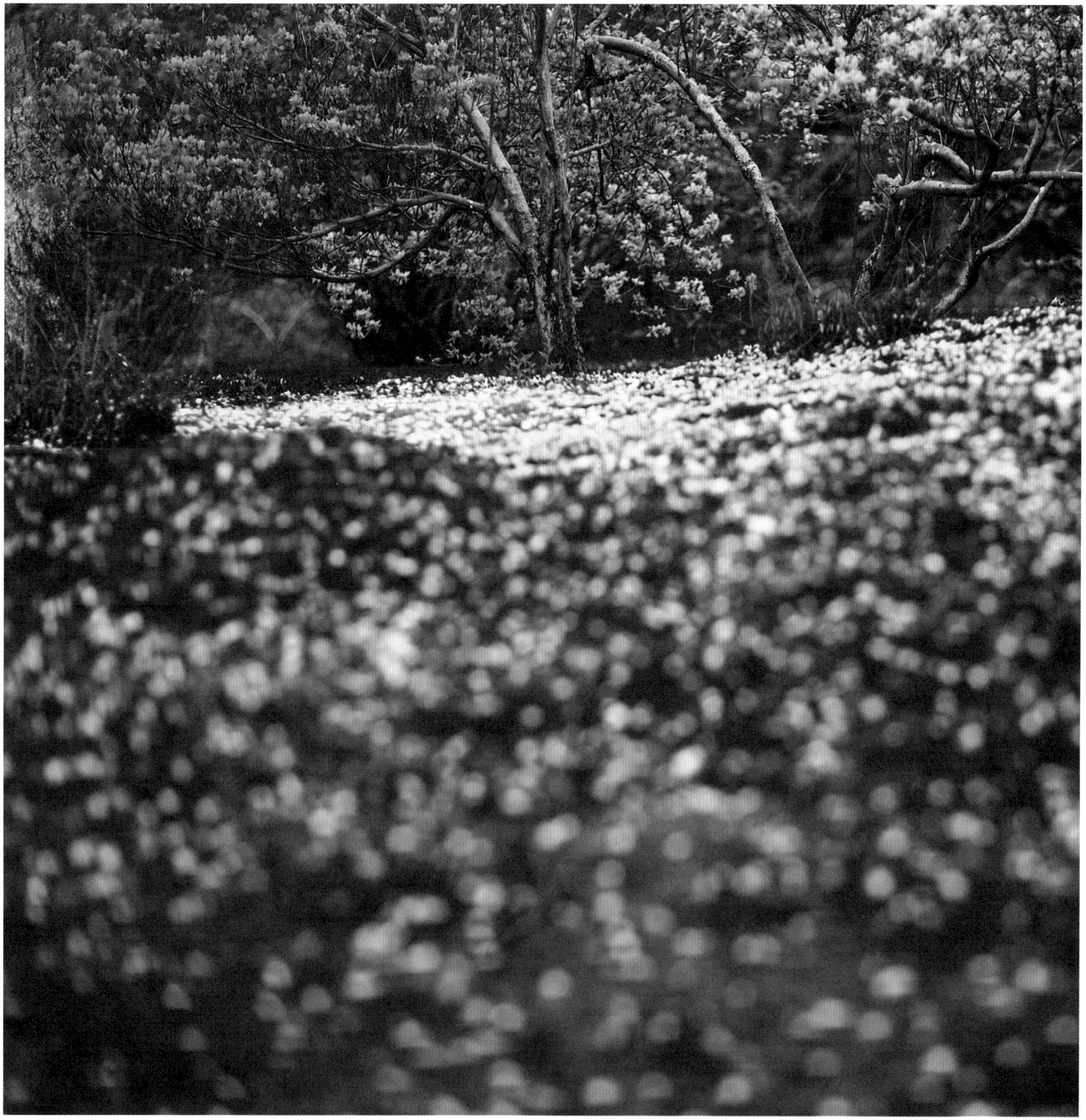

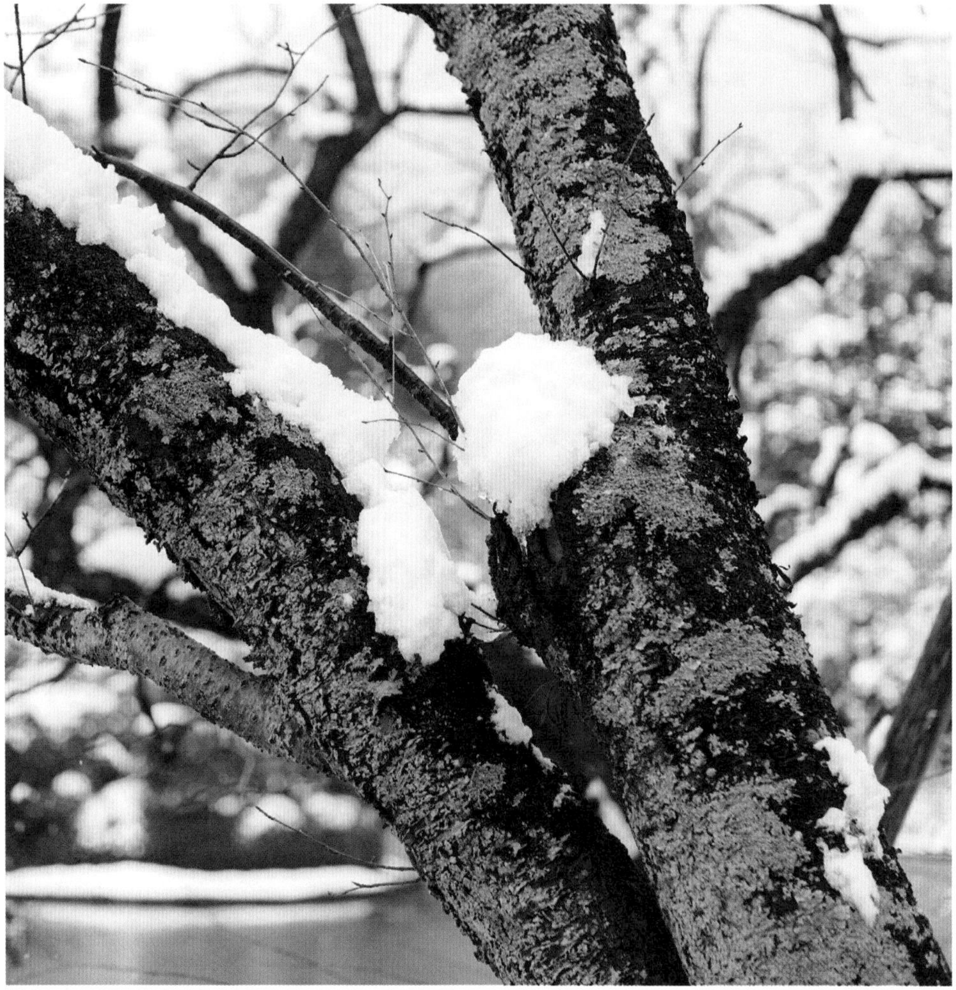

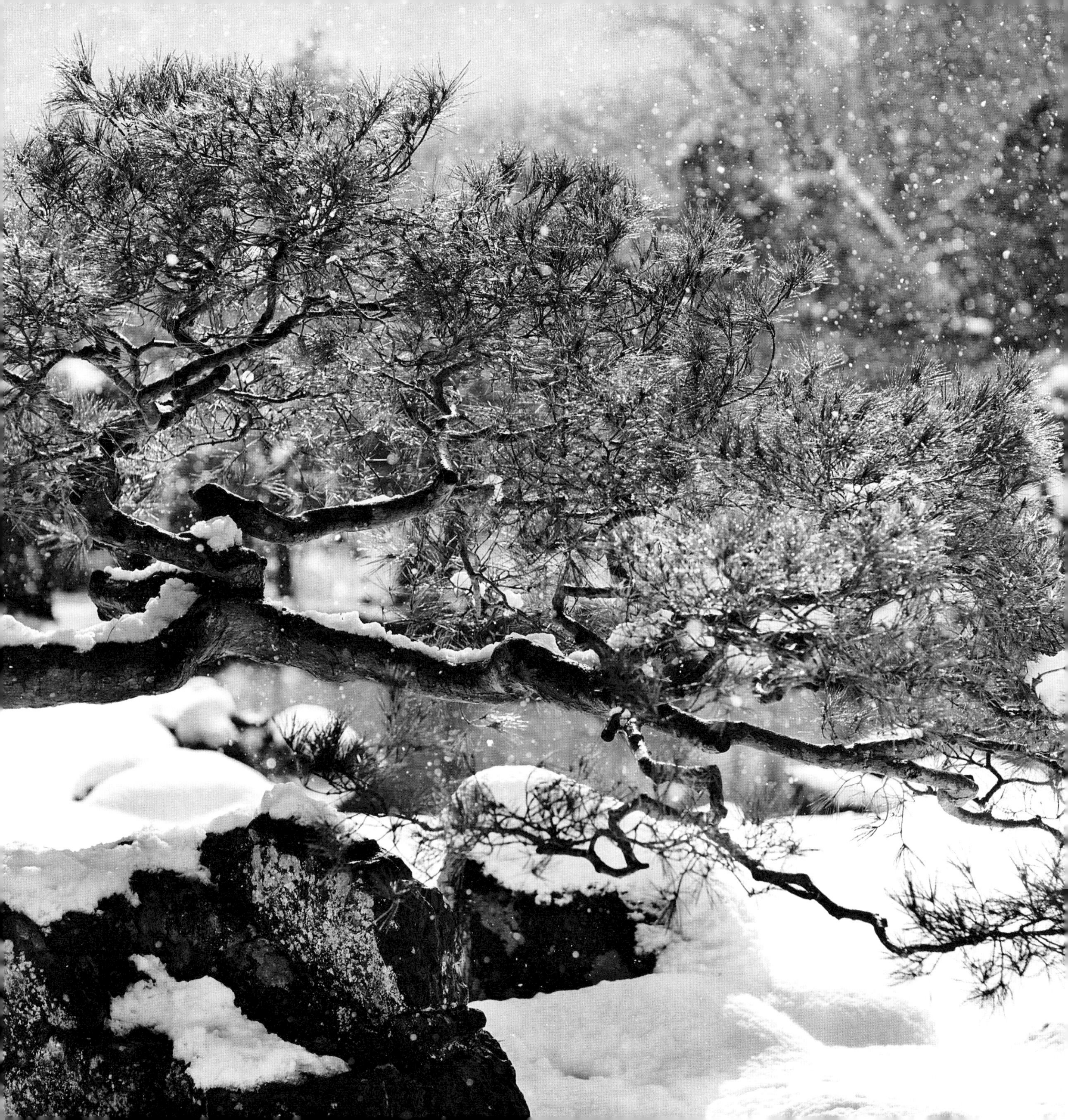

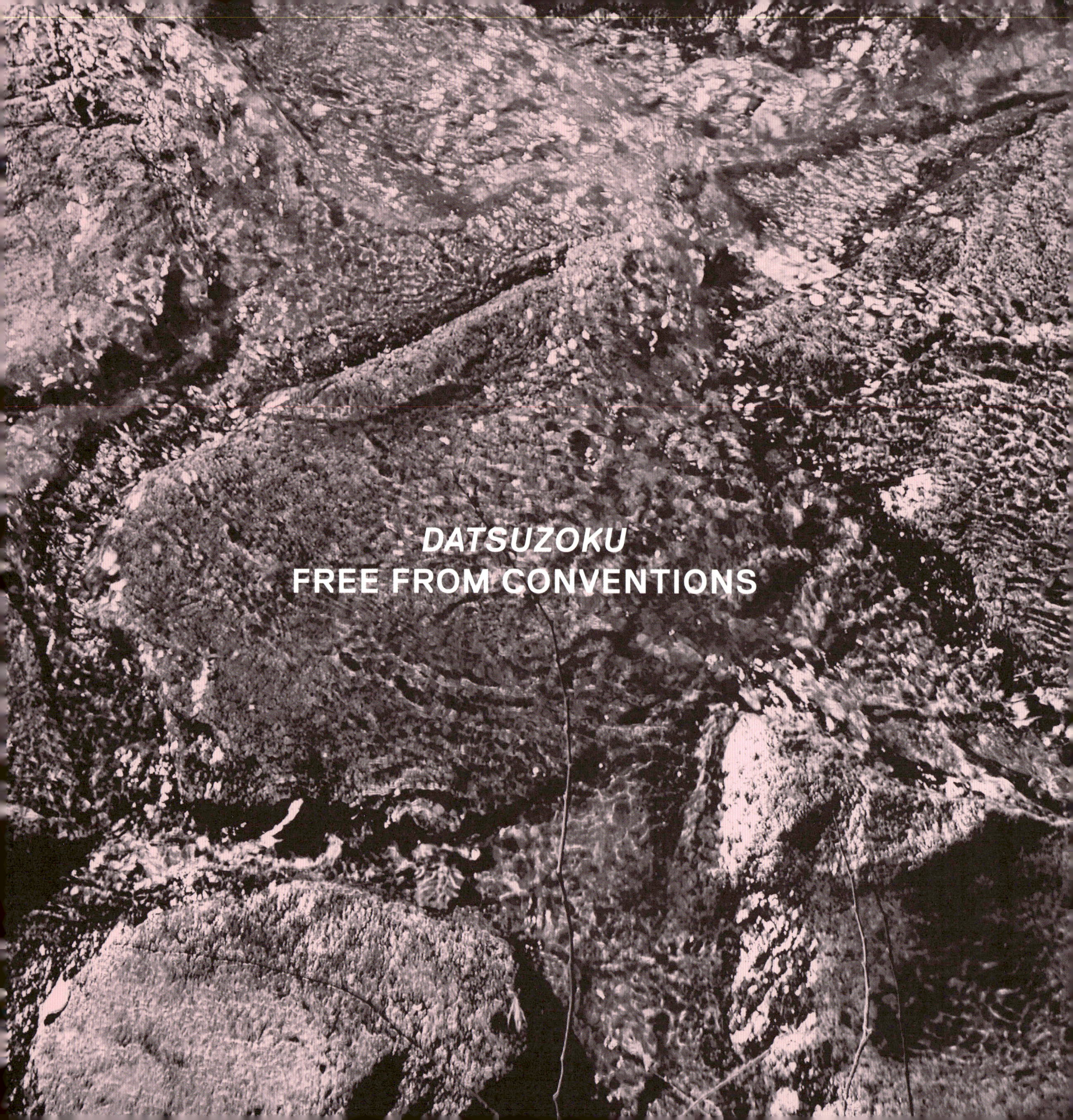
DATSUZOKU
FREE FROM CONVENTIONS

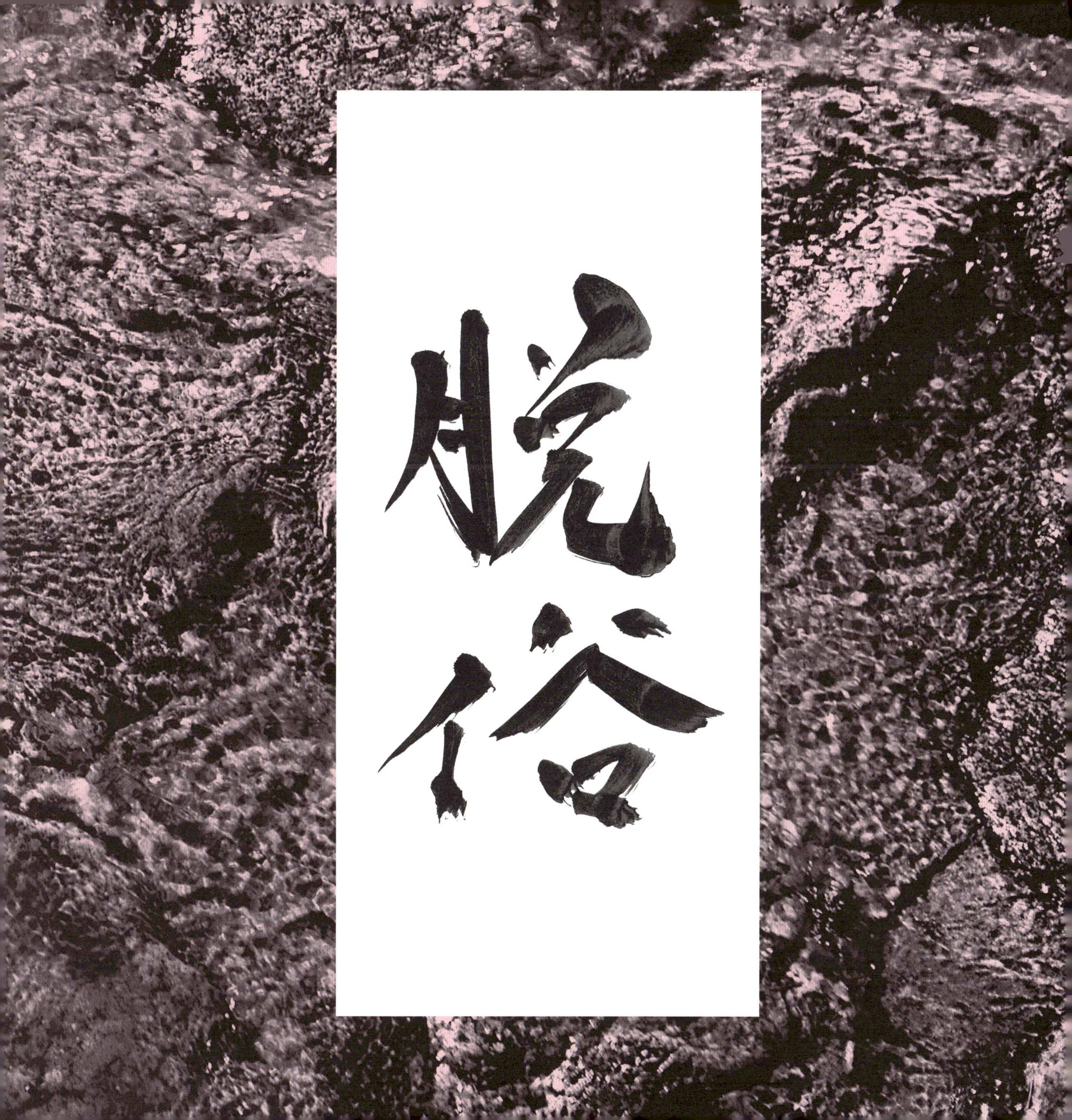
脱俗

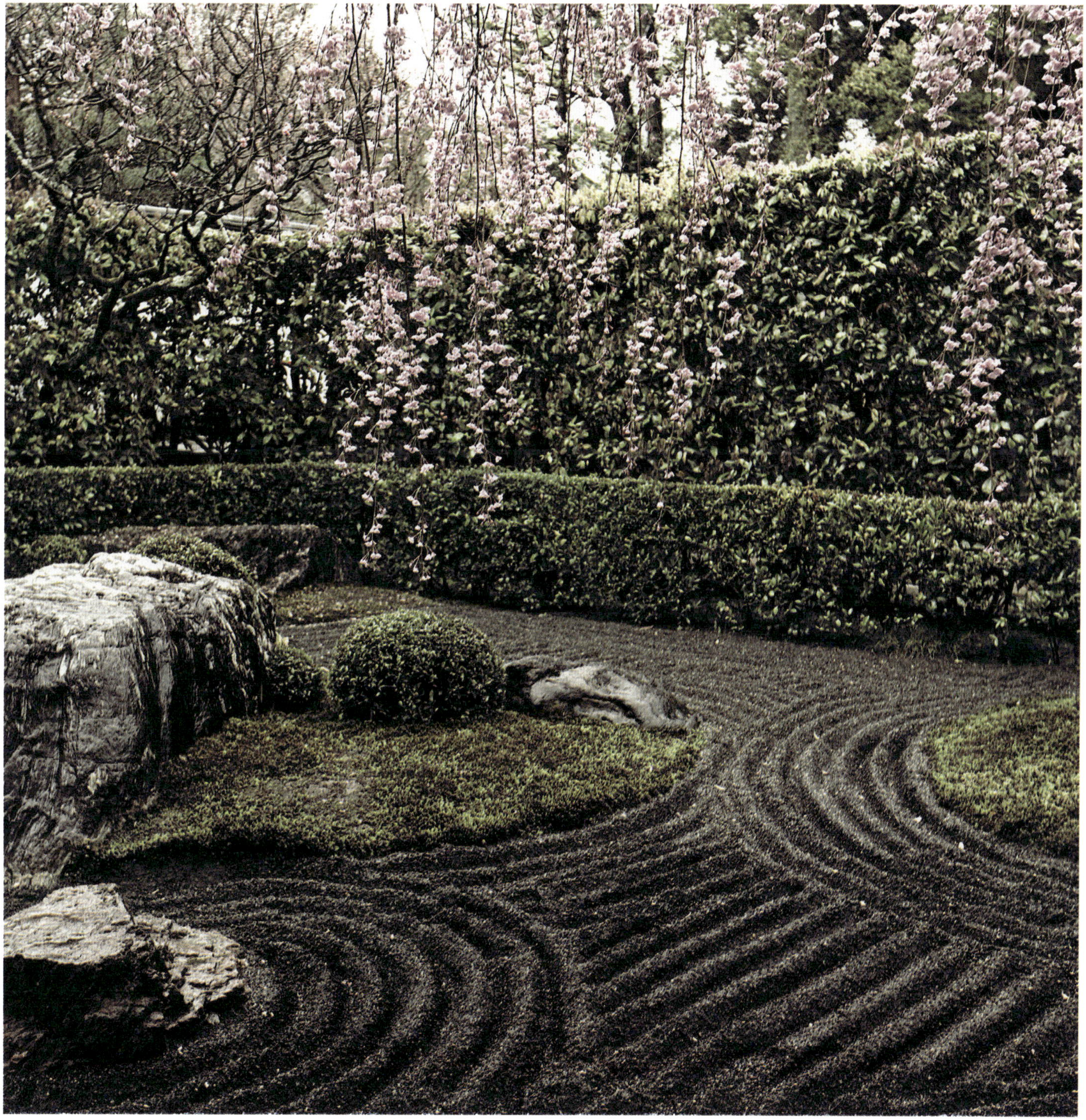

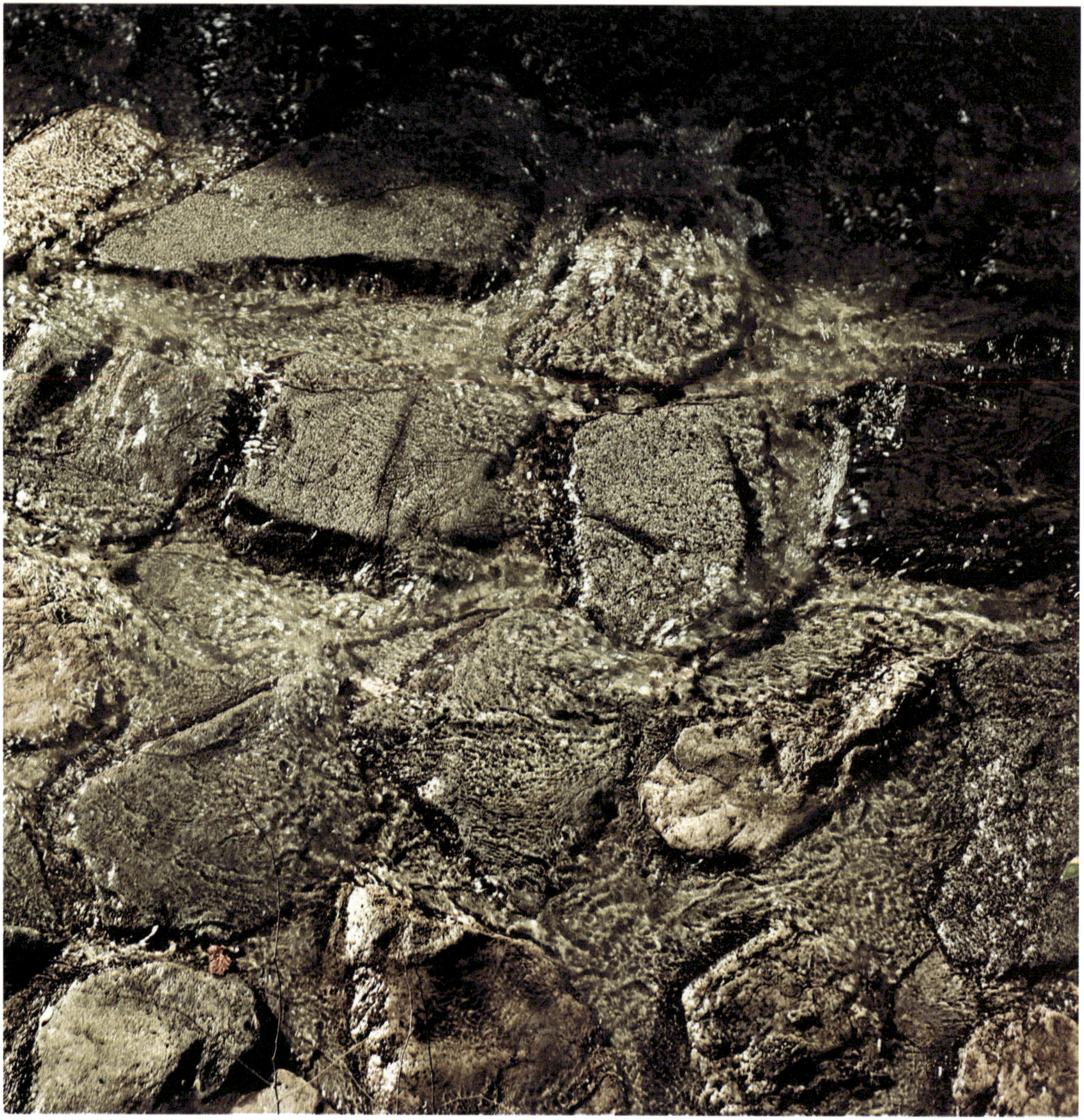

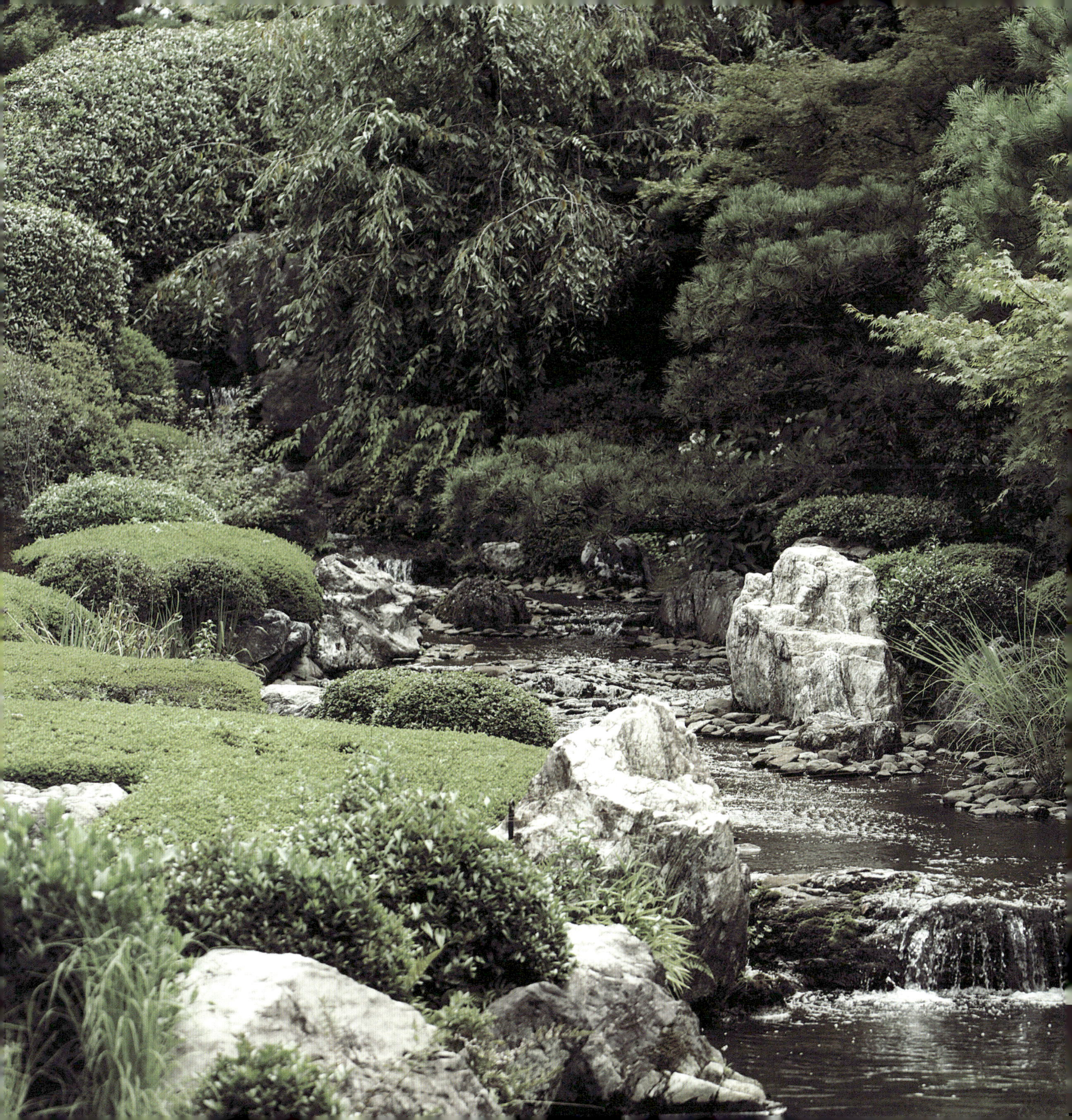

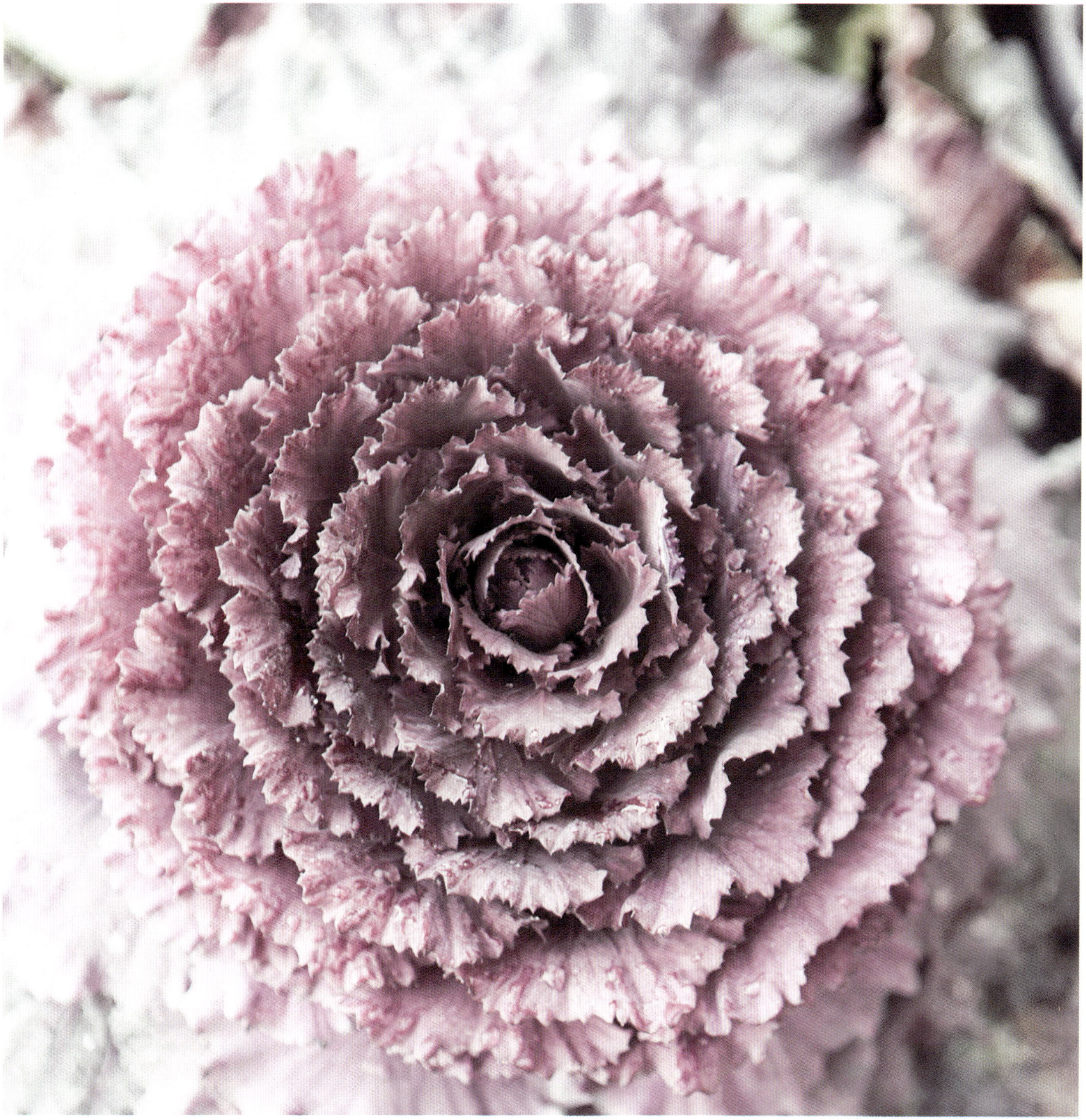

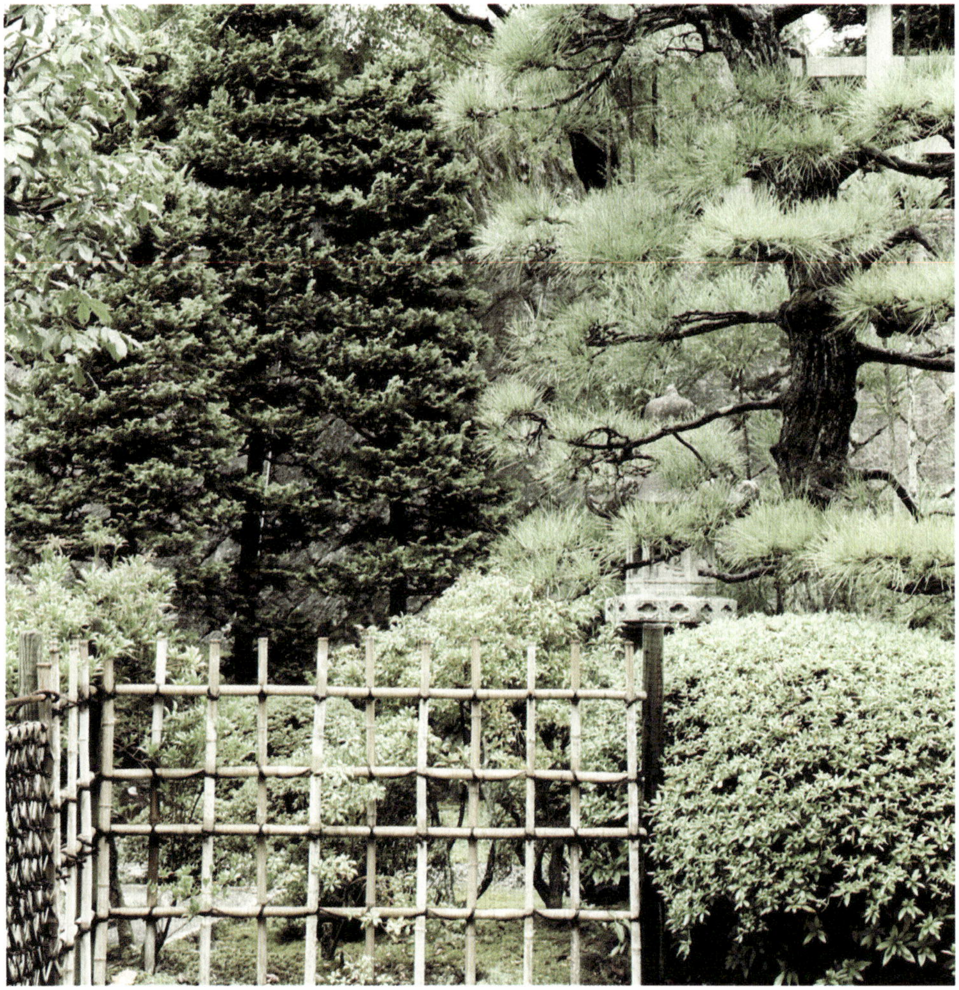

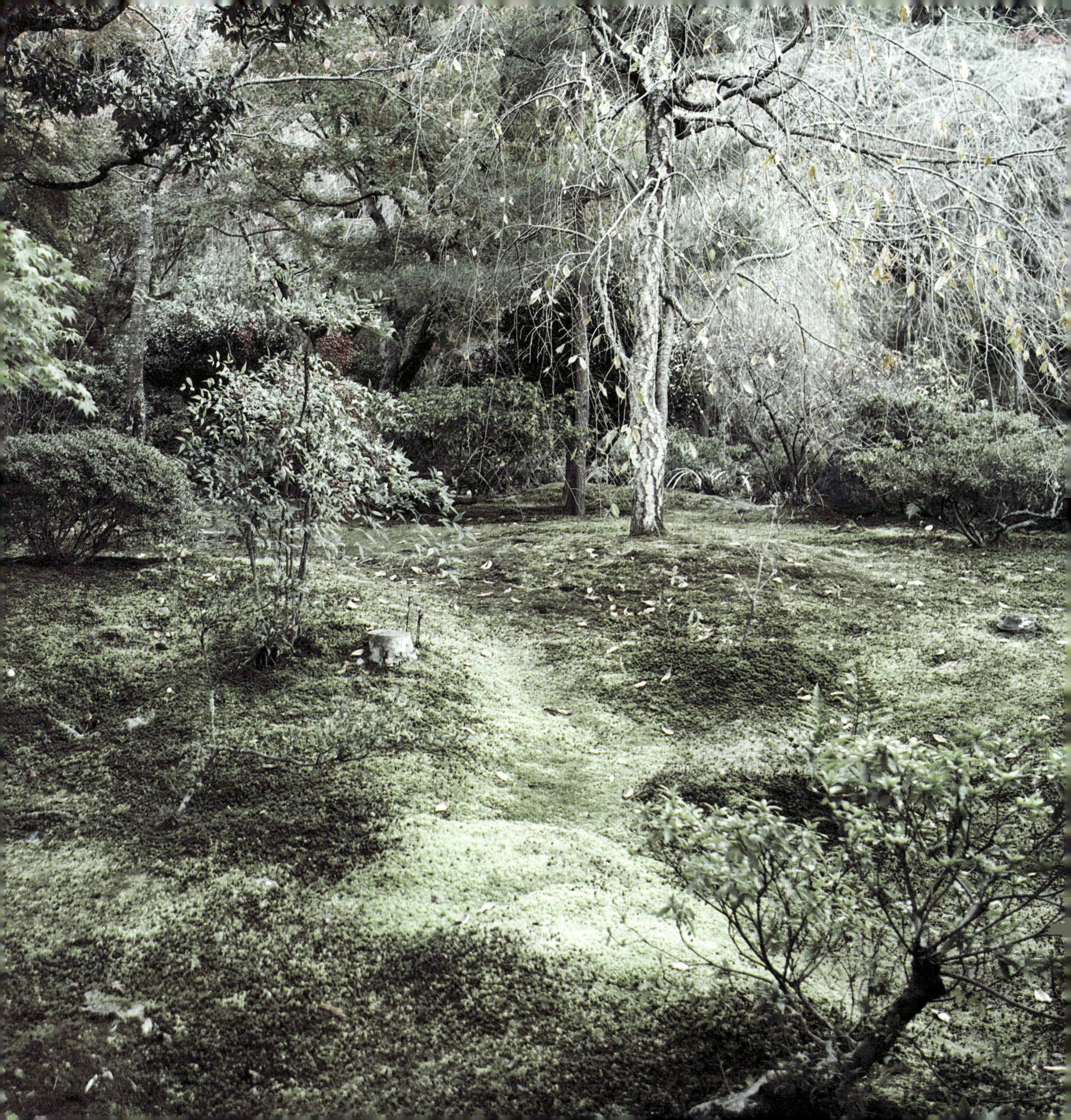

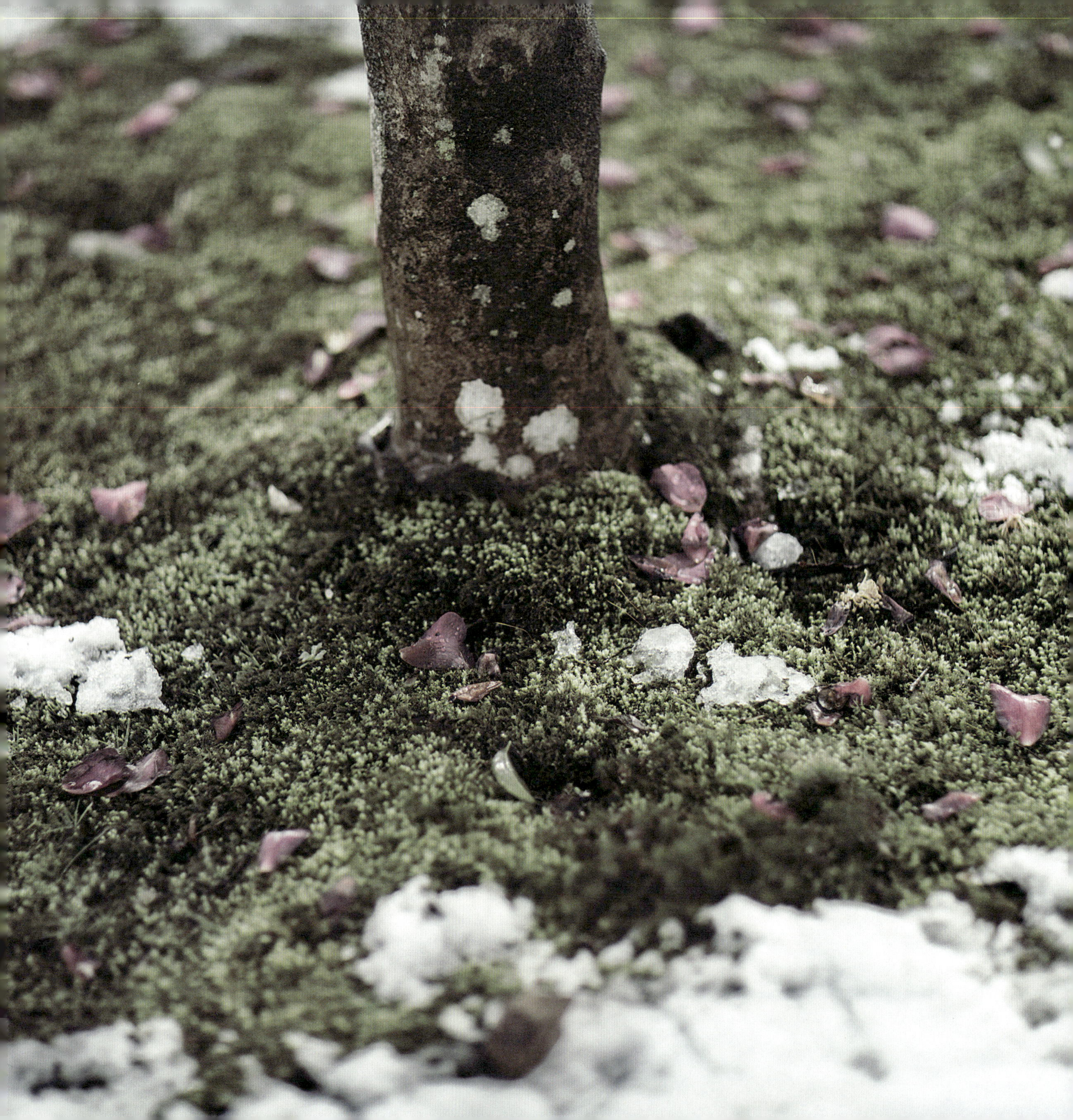

THE SECRET OF LIFE IS IN THE HEART.

Ryōkan

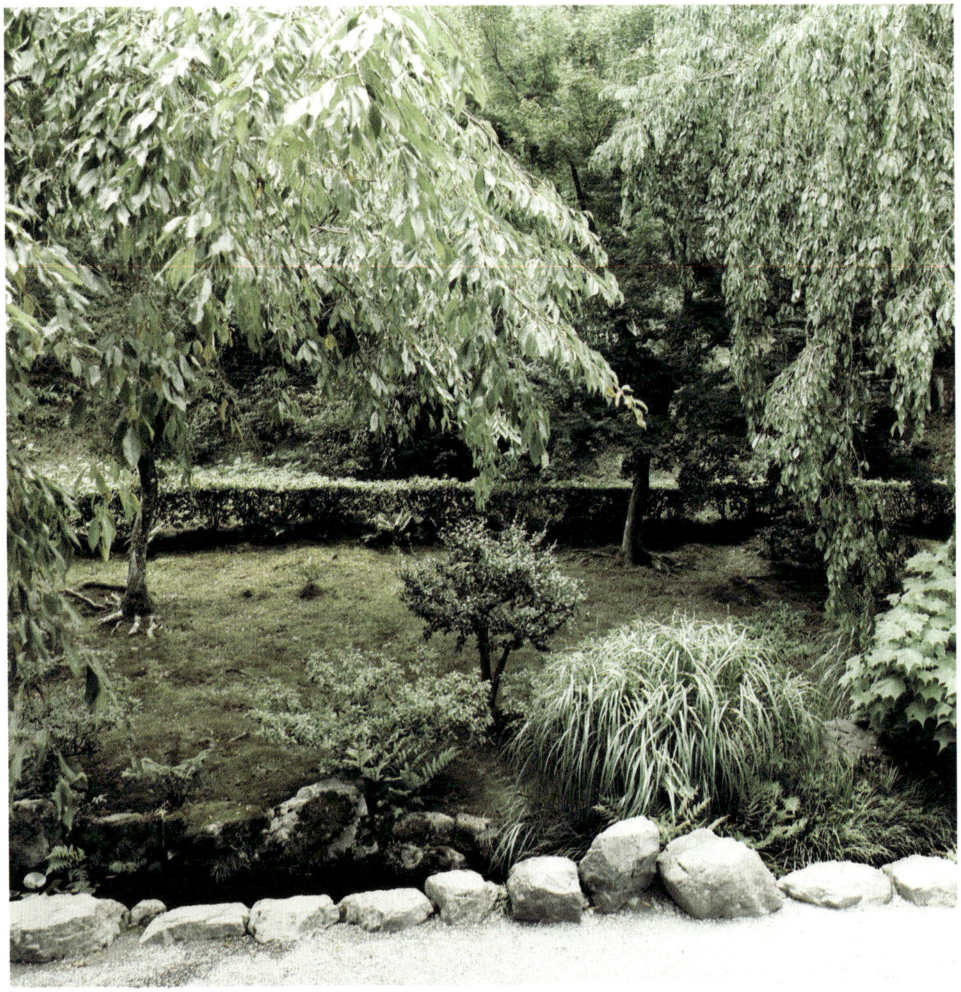

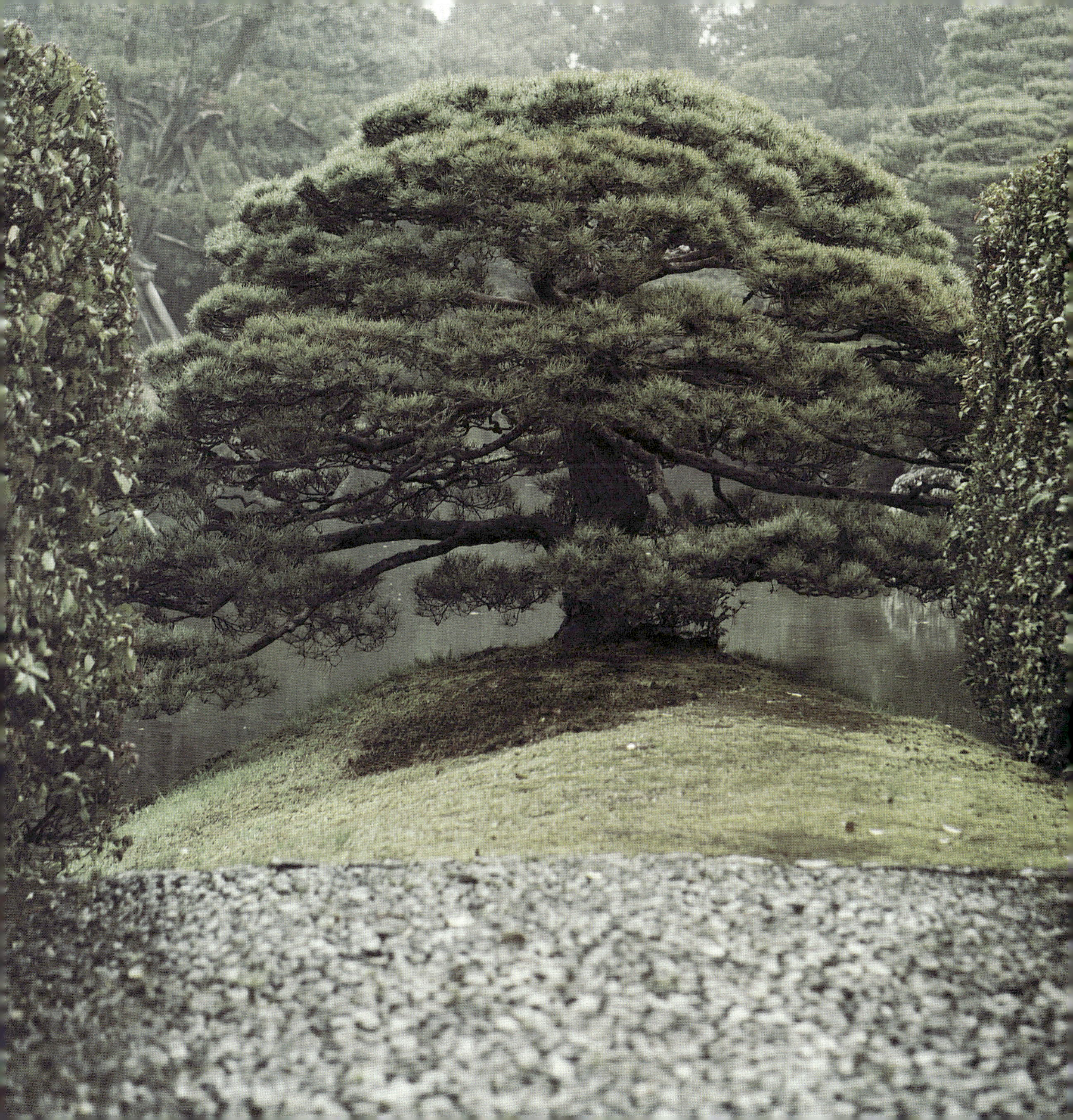

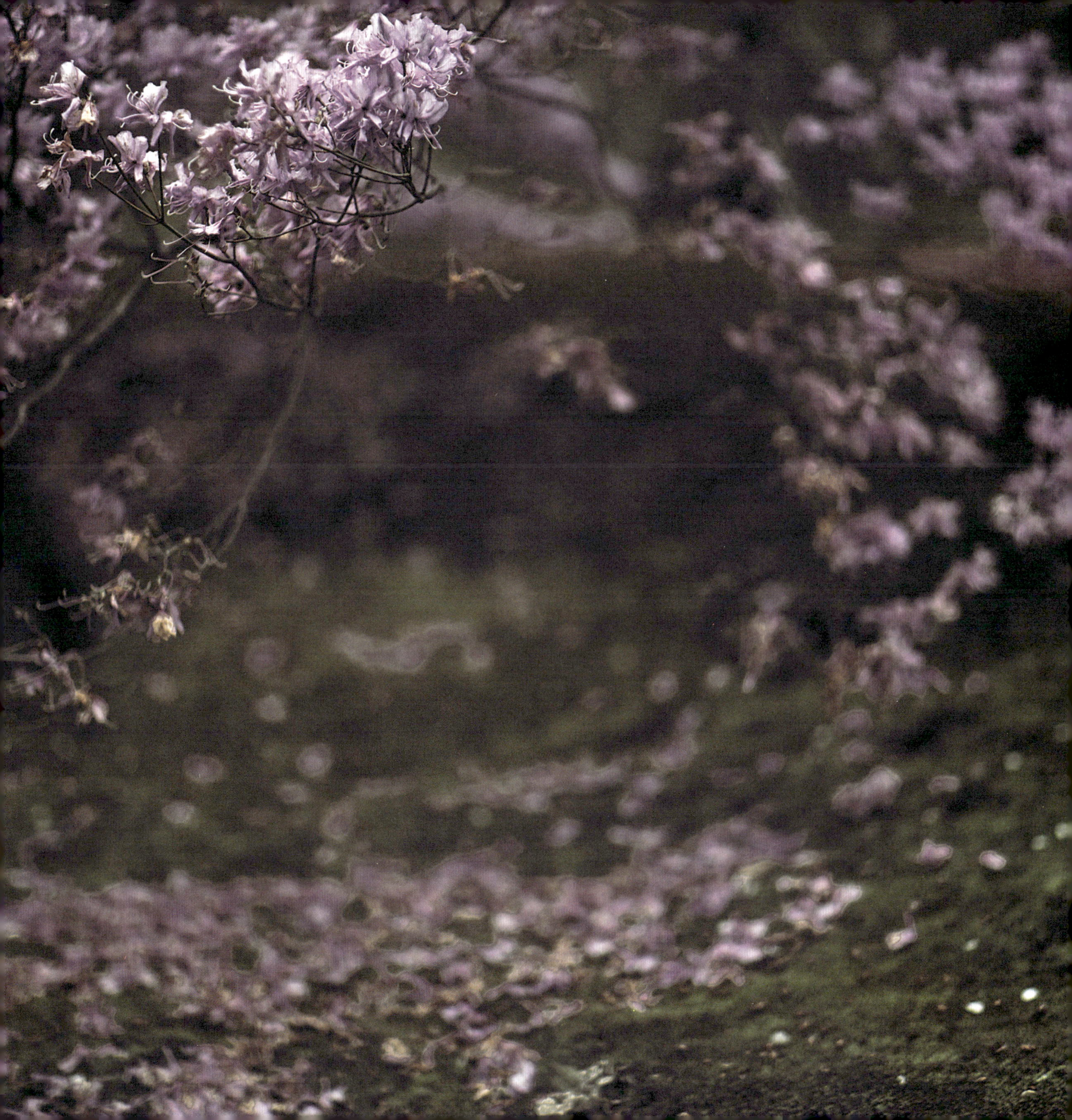

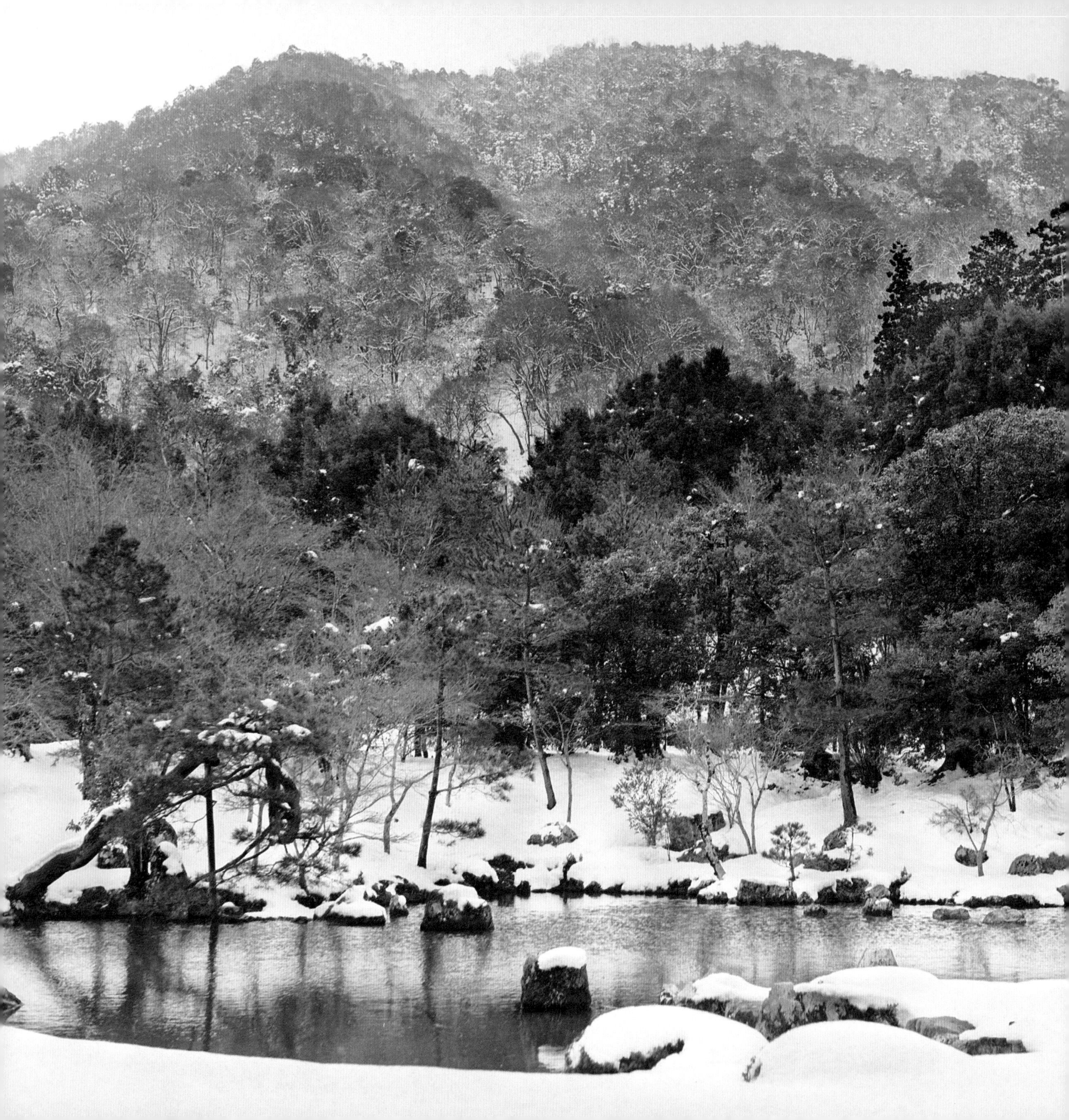

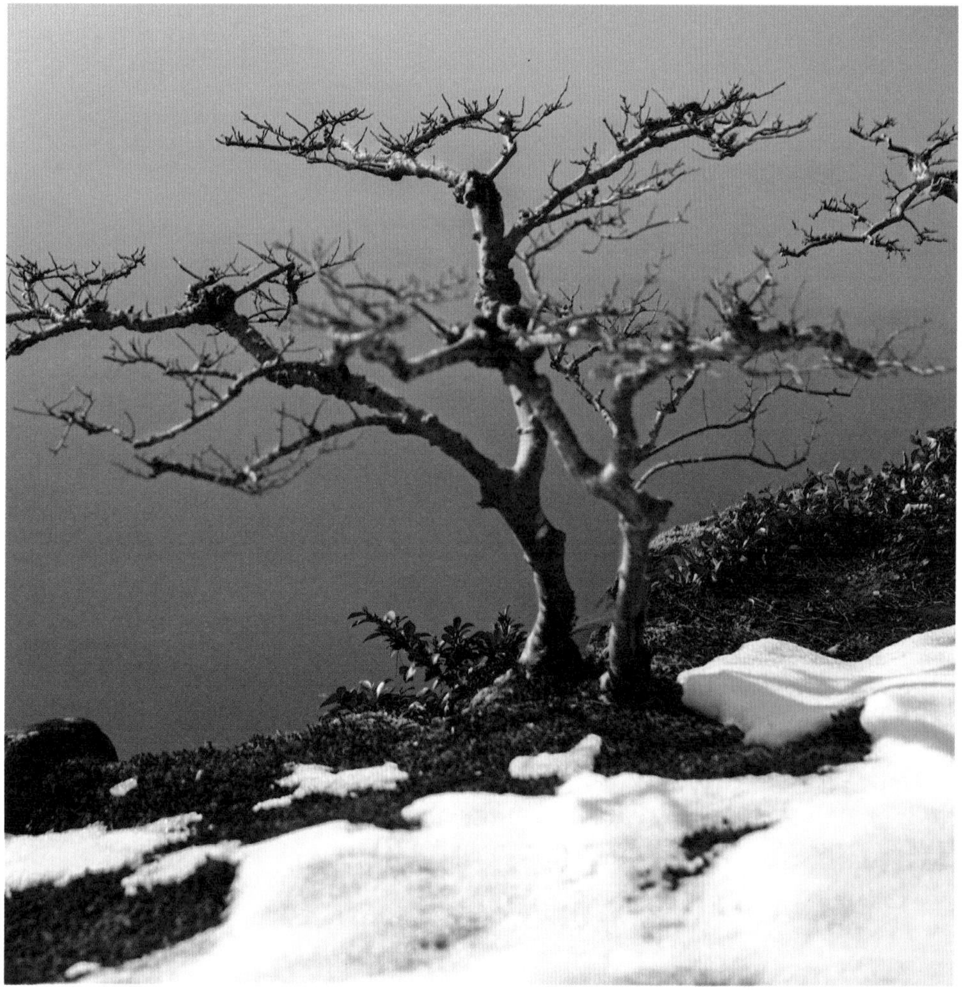

A DAY WITHOUT WORKING
IS A DAY WITHOUT EATING.

Baizhang Huaihai

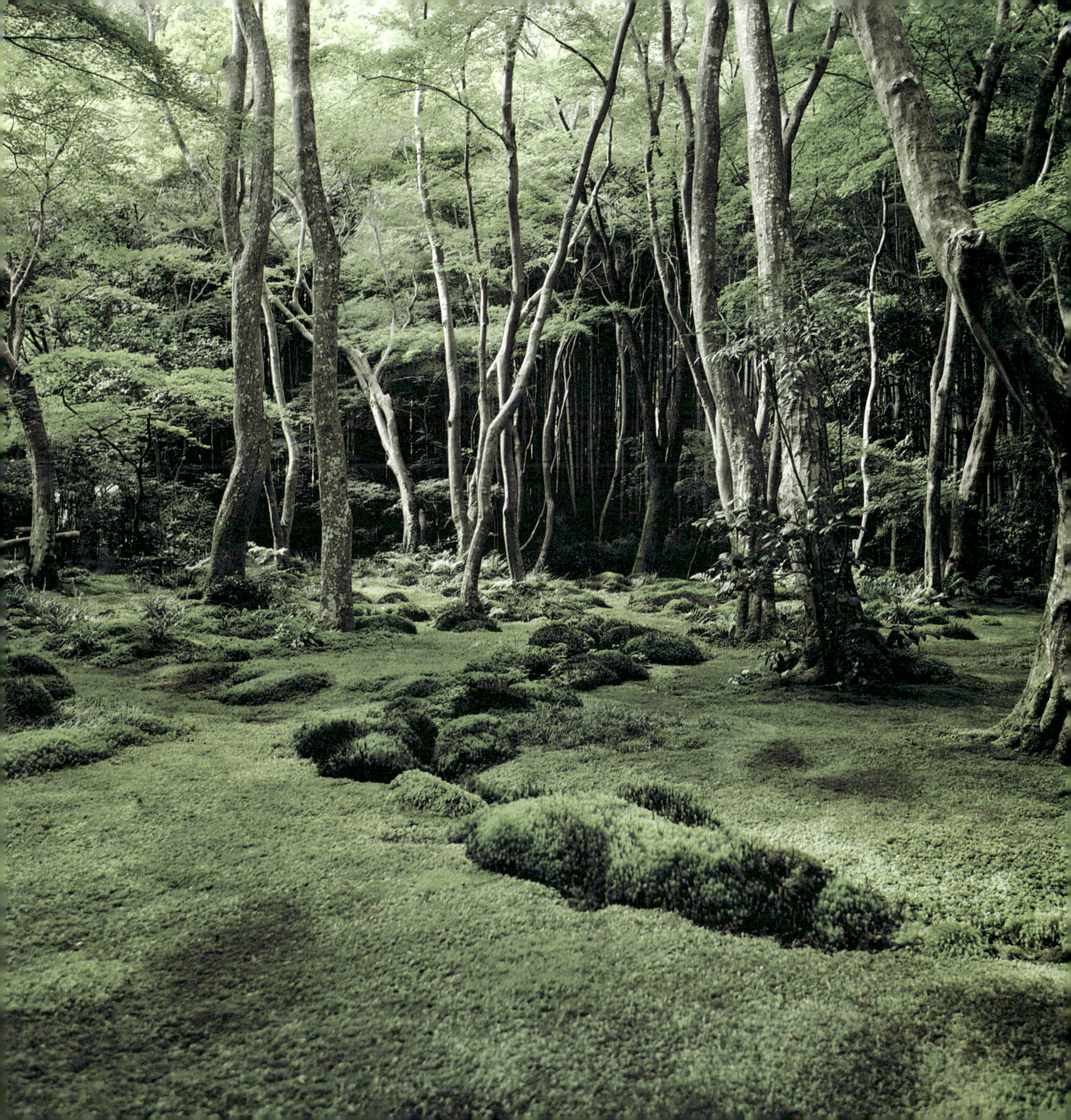

SEIJAKU TRANQUILITY

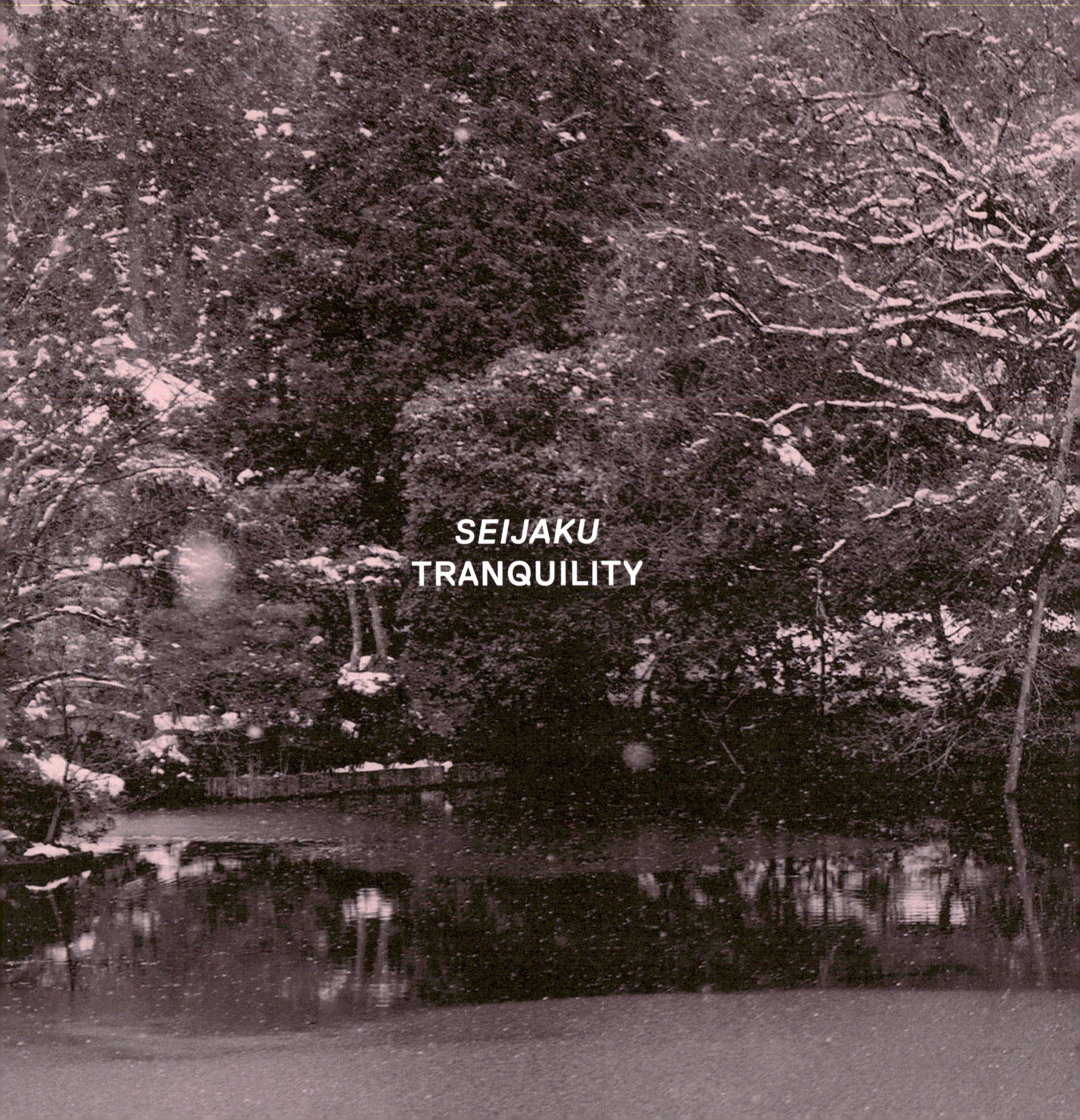

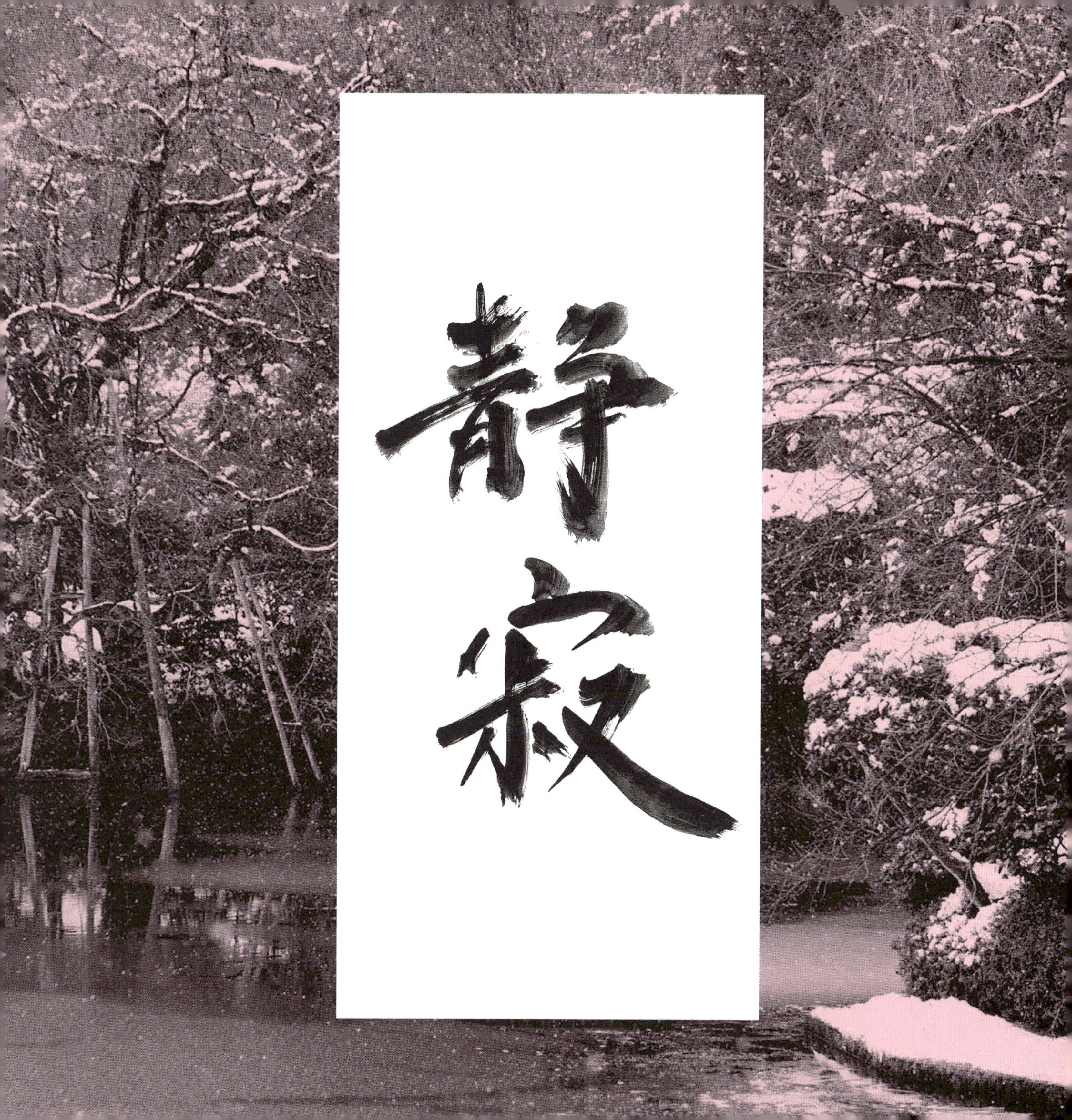
静寂

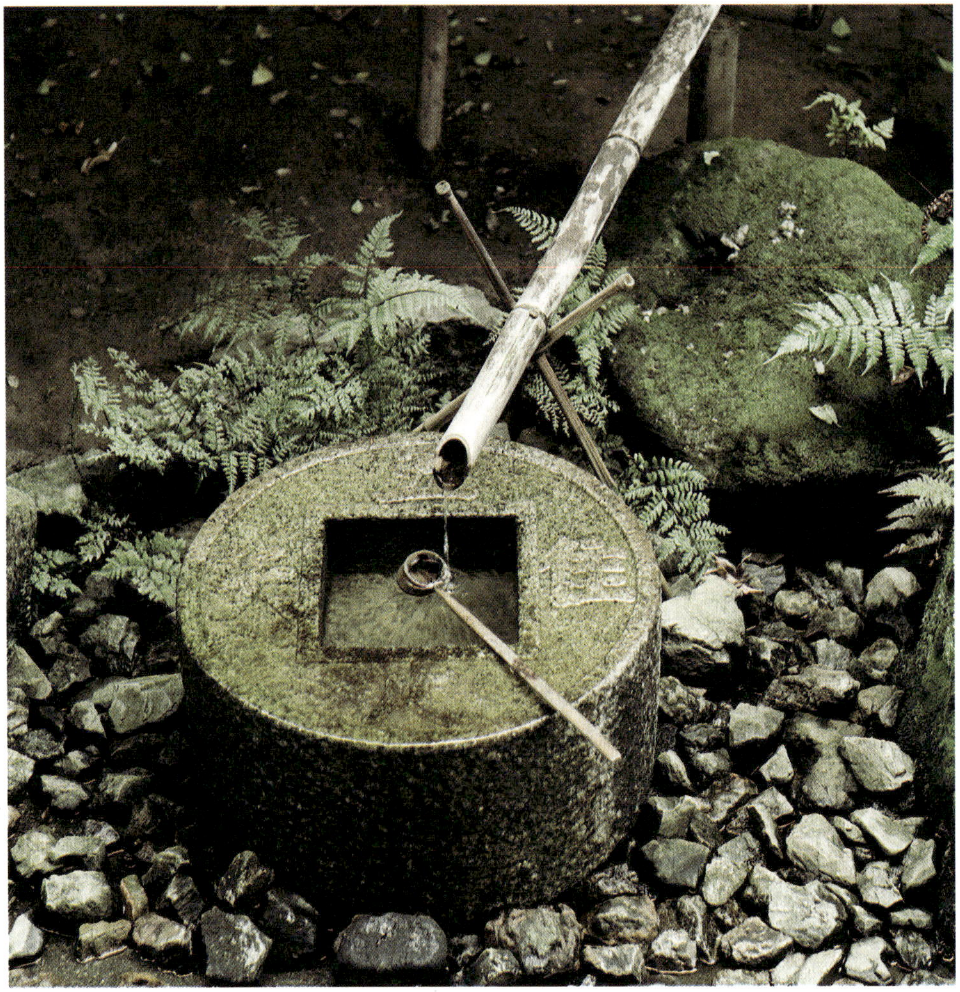

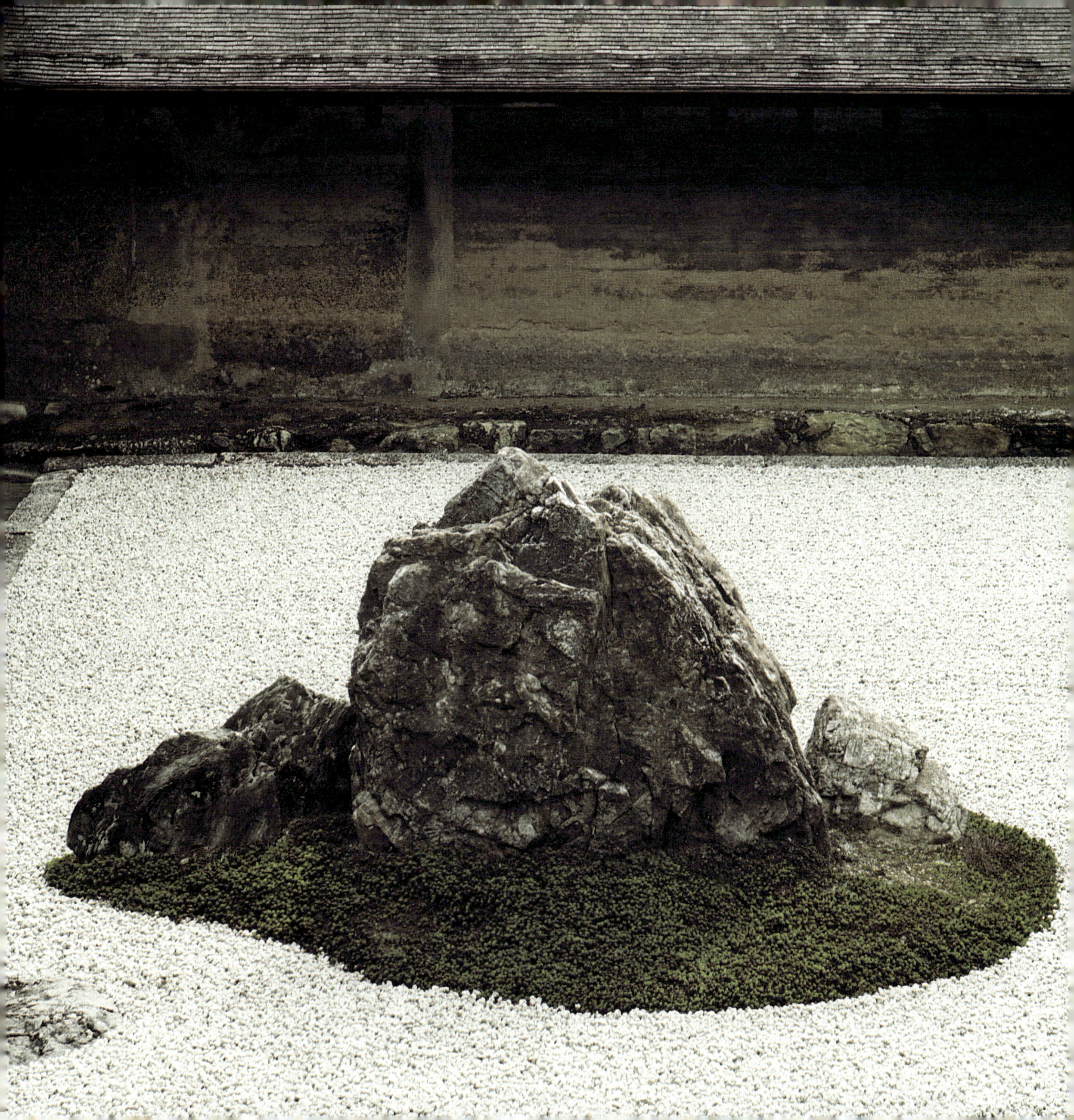

HAPPINESS IS HERE AND NOW.

Thich Nhat Hanh

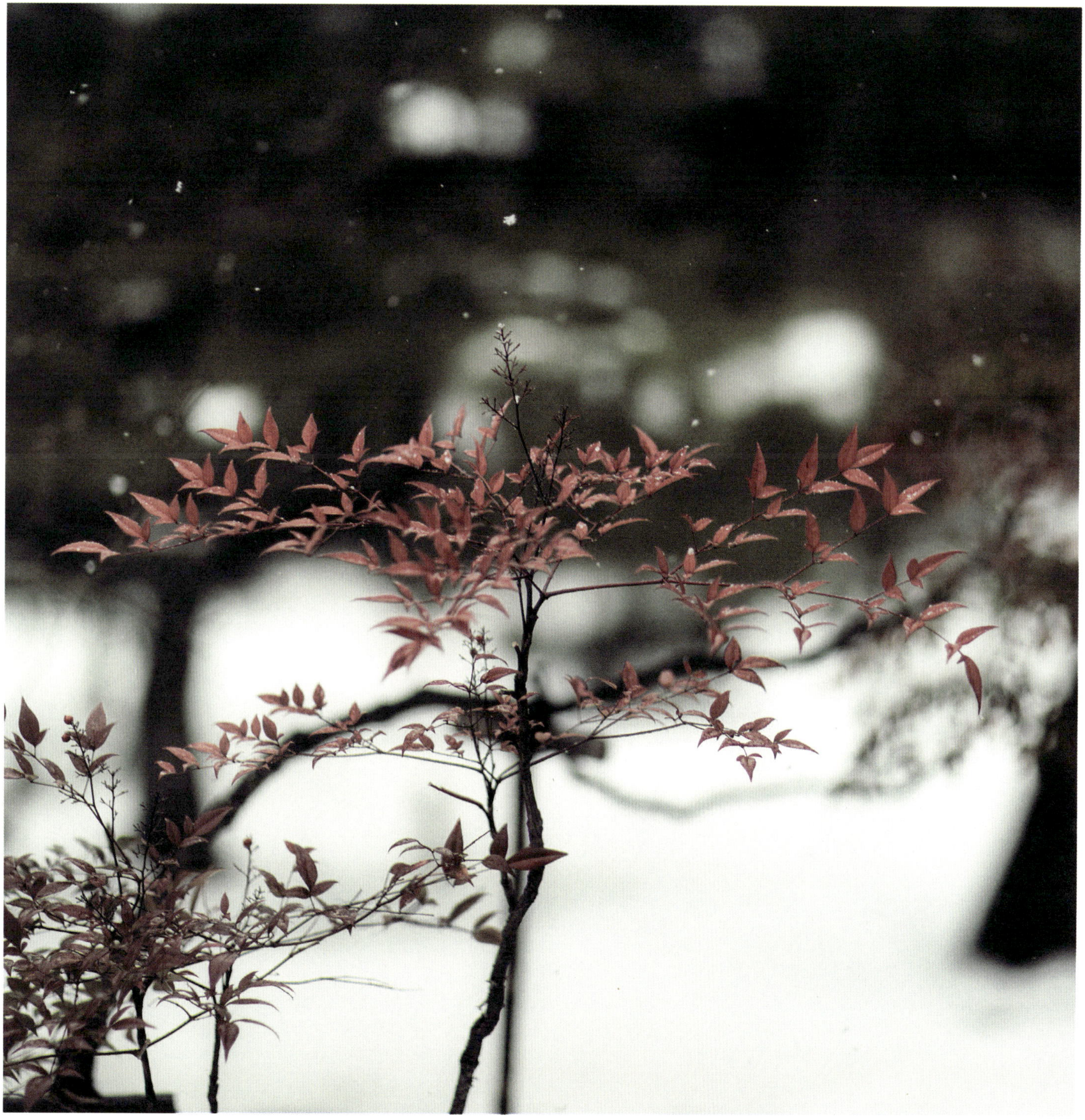

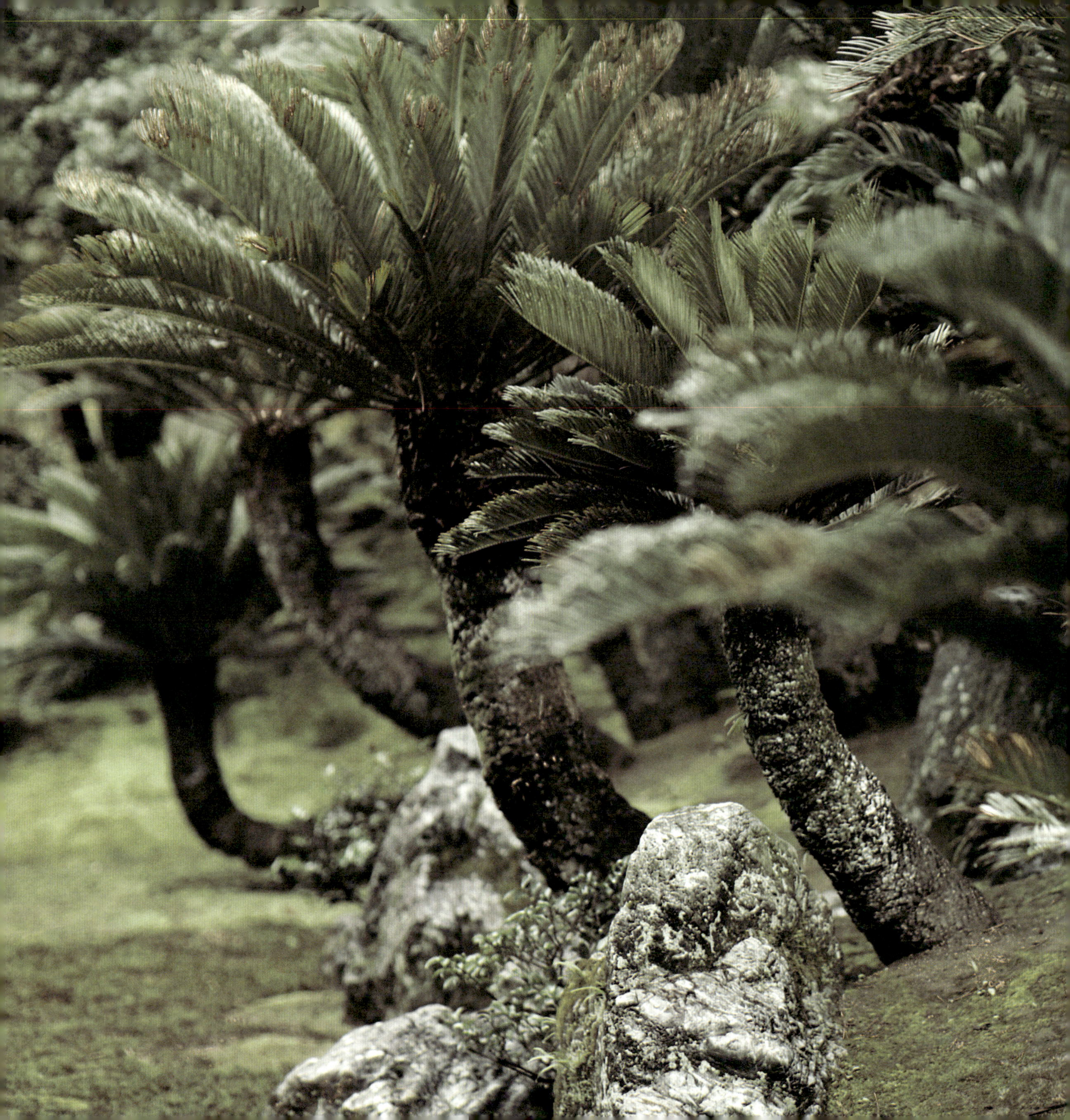

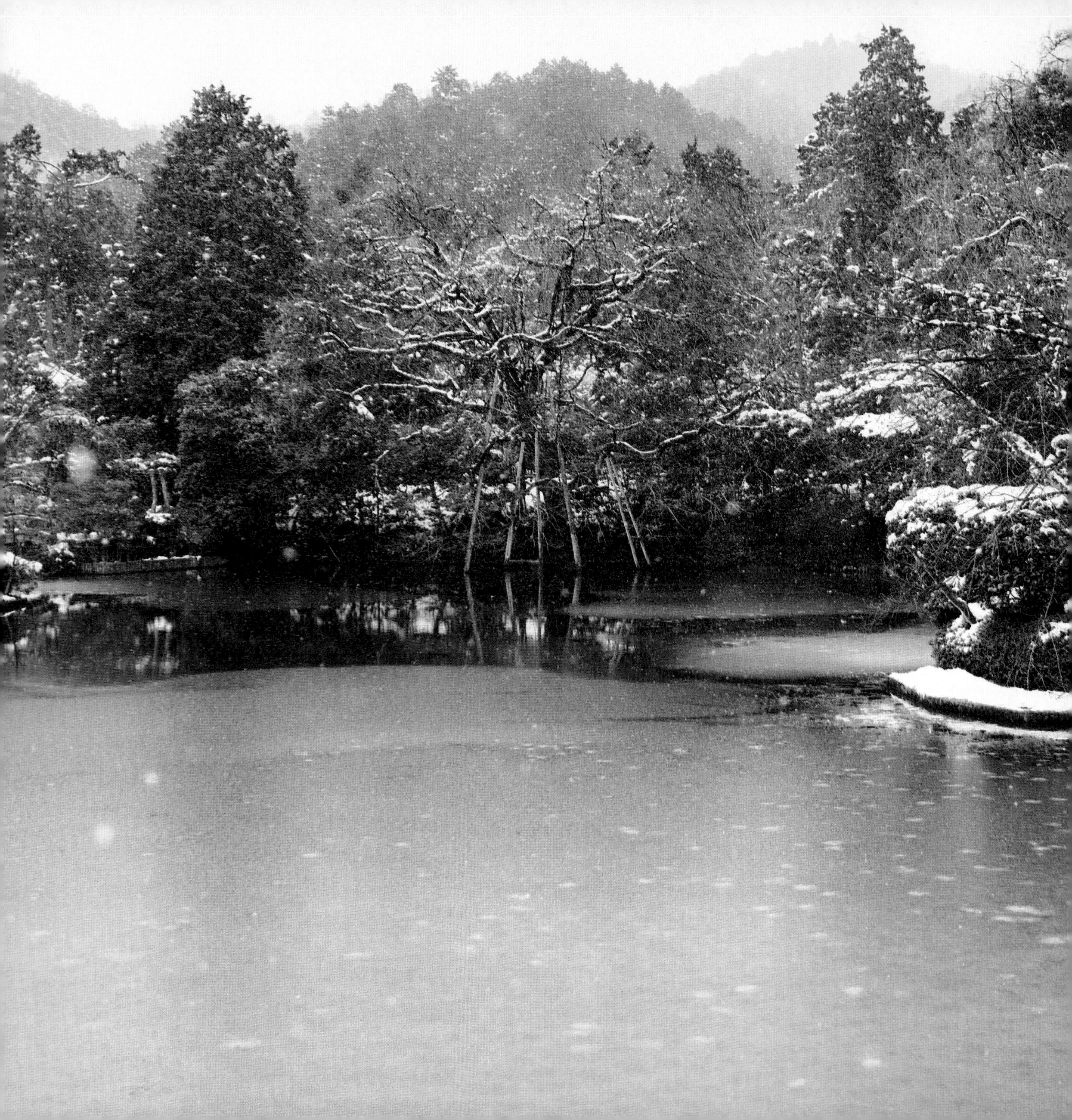

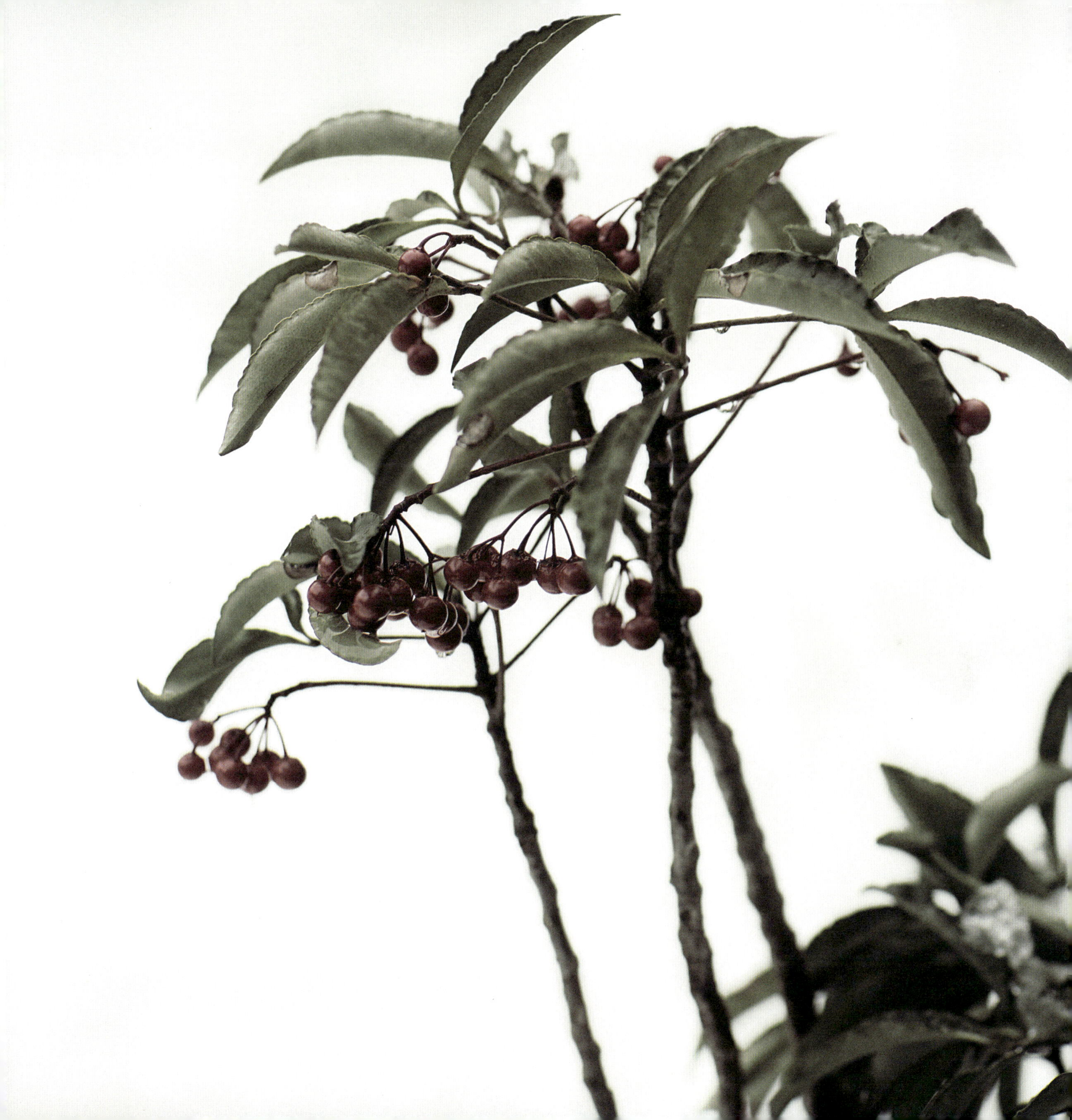

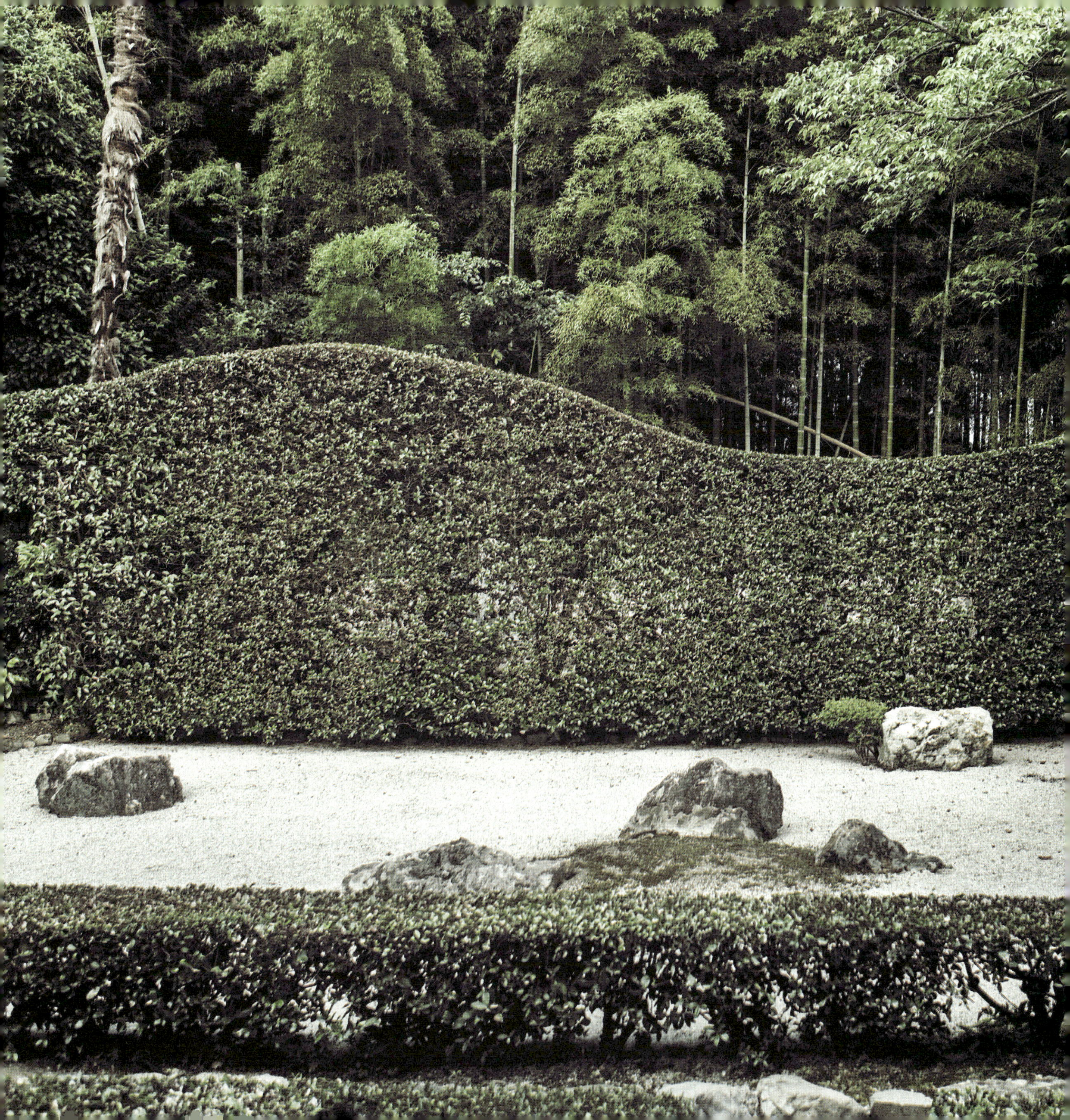

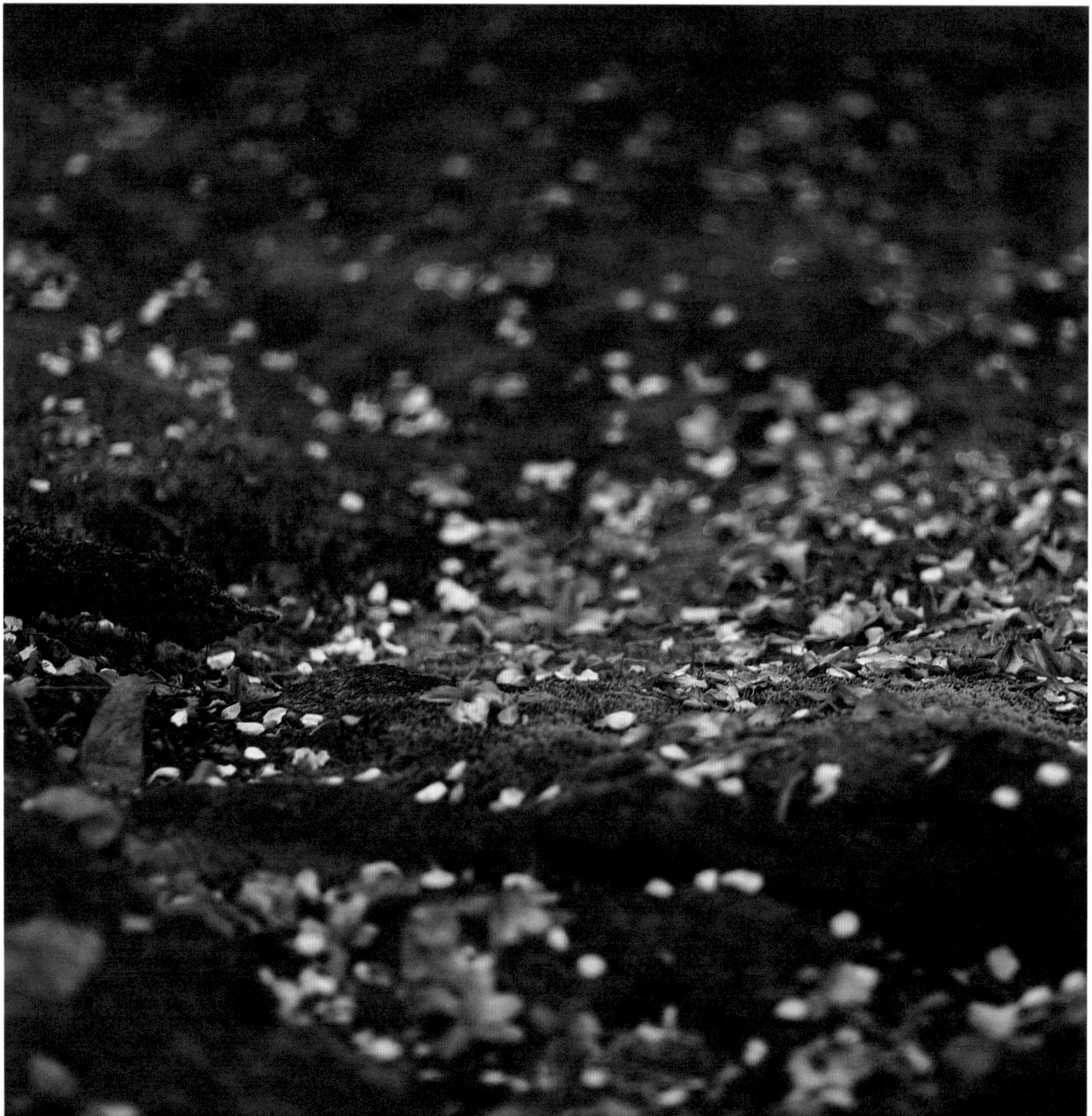

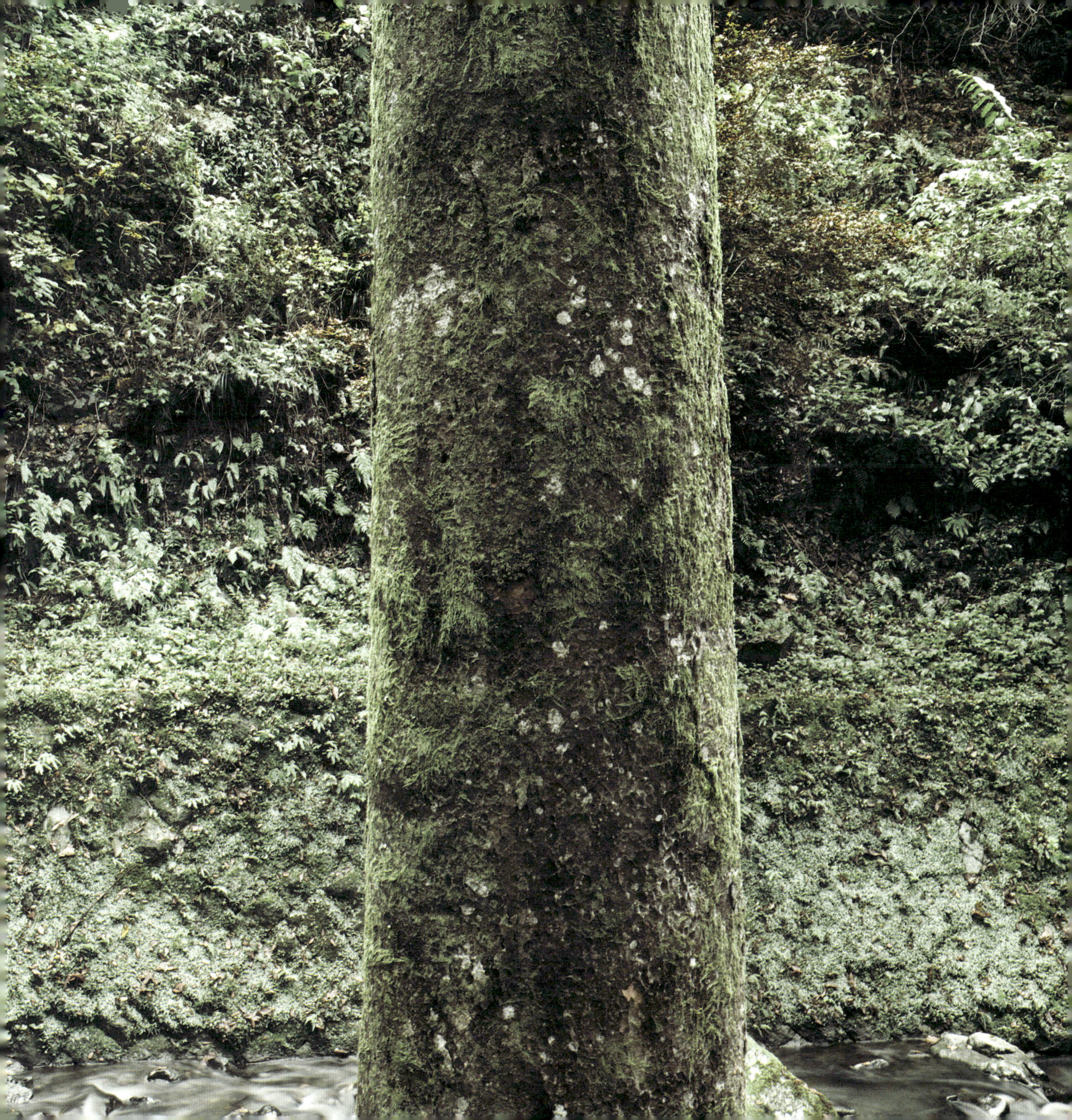

SITTING SILENTLY, DOING NOTHING,
SPRING COMES, AND THE GRASS GROWS BY ITSELF.

Unknown

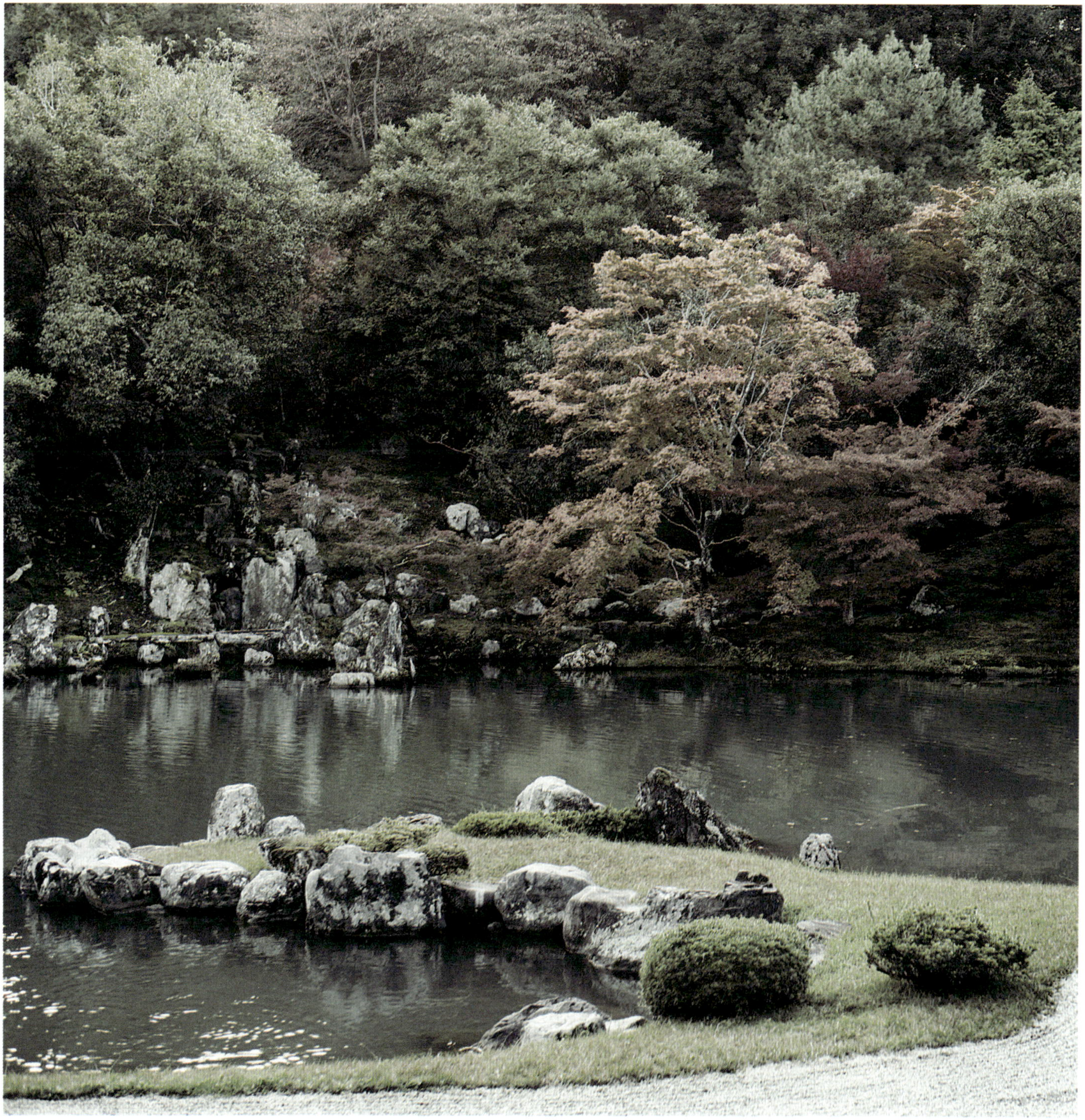

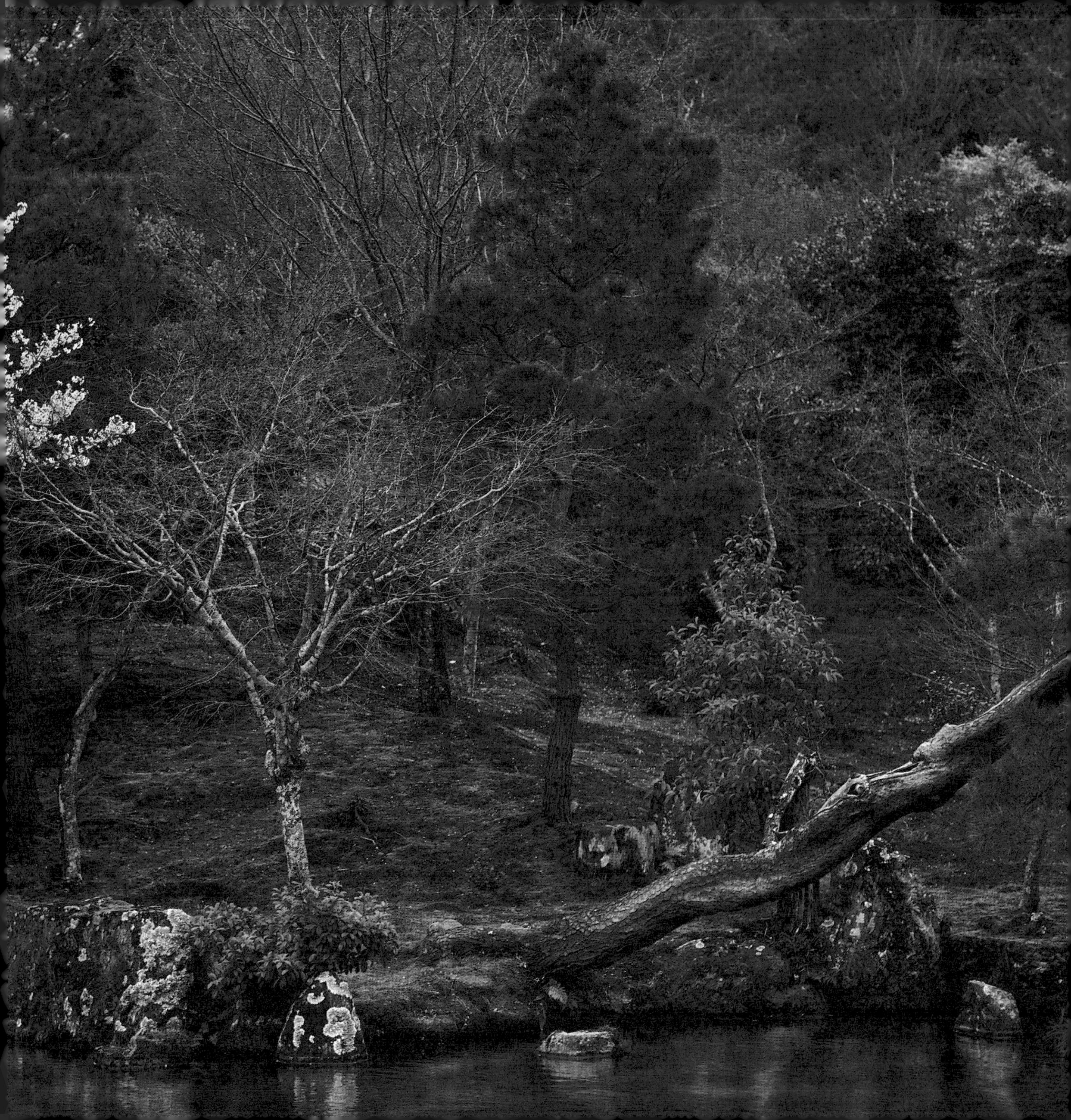

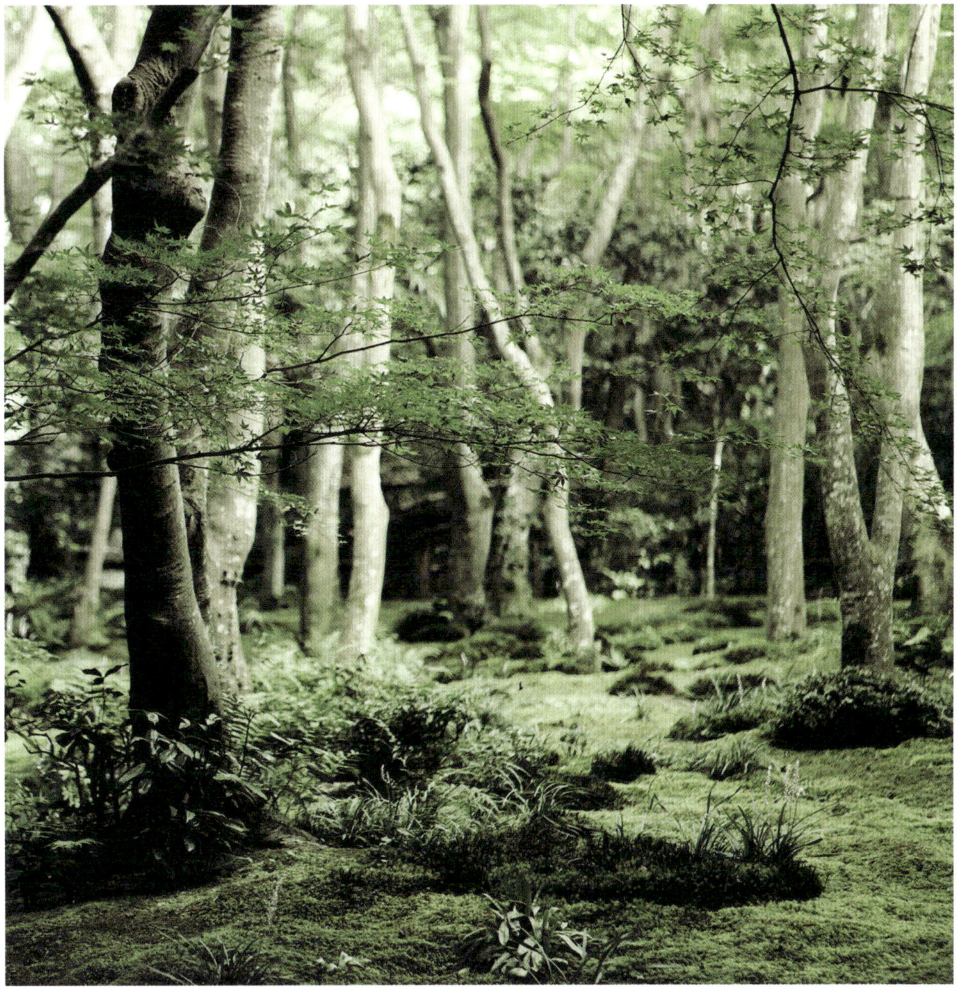

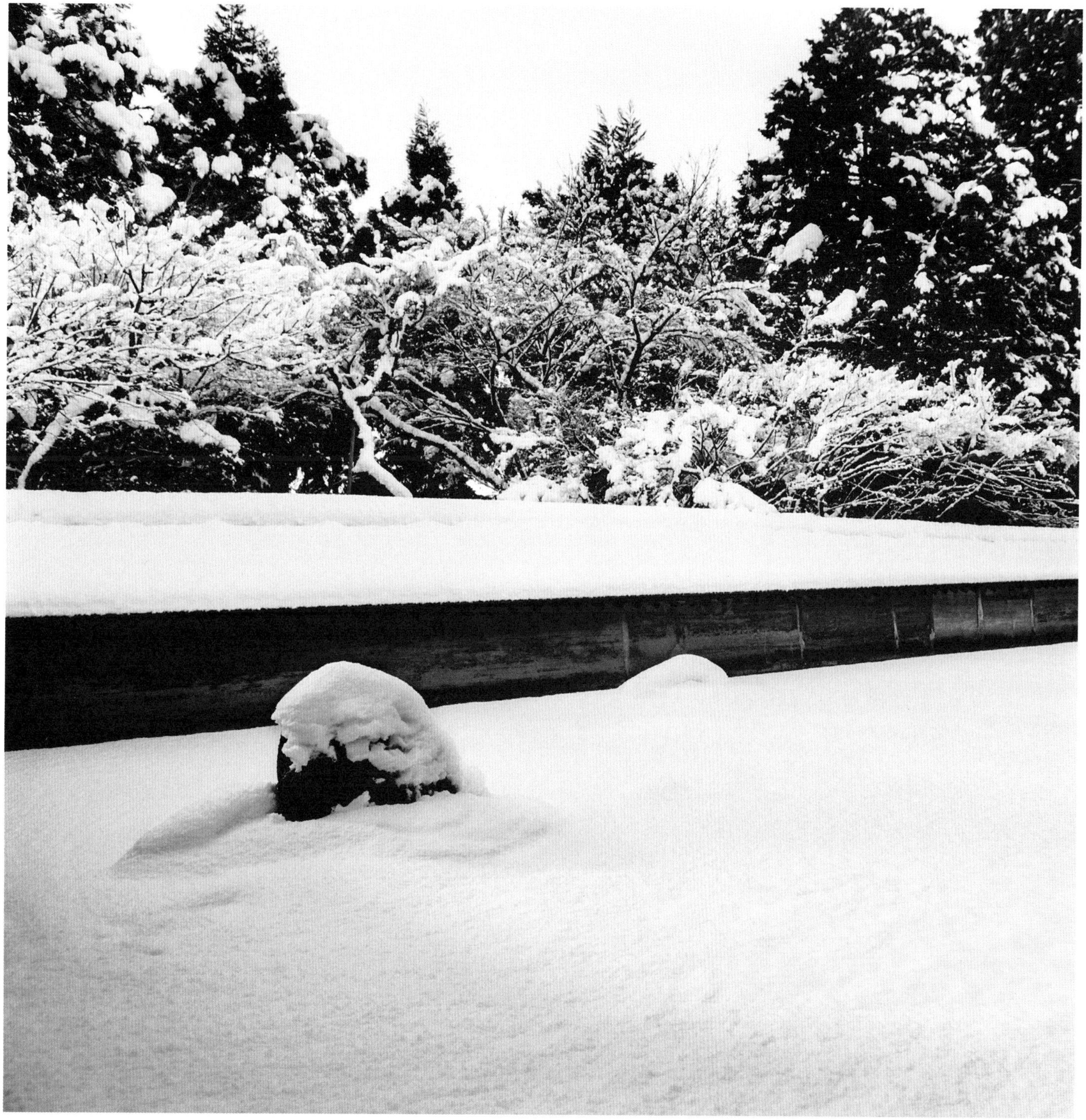

Dedicated to
Kolja and Mika

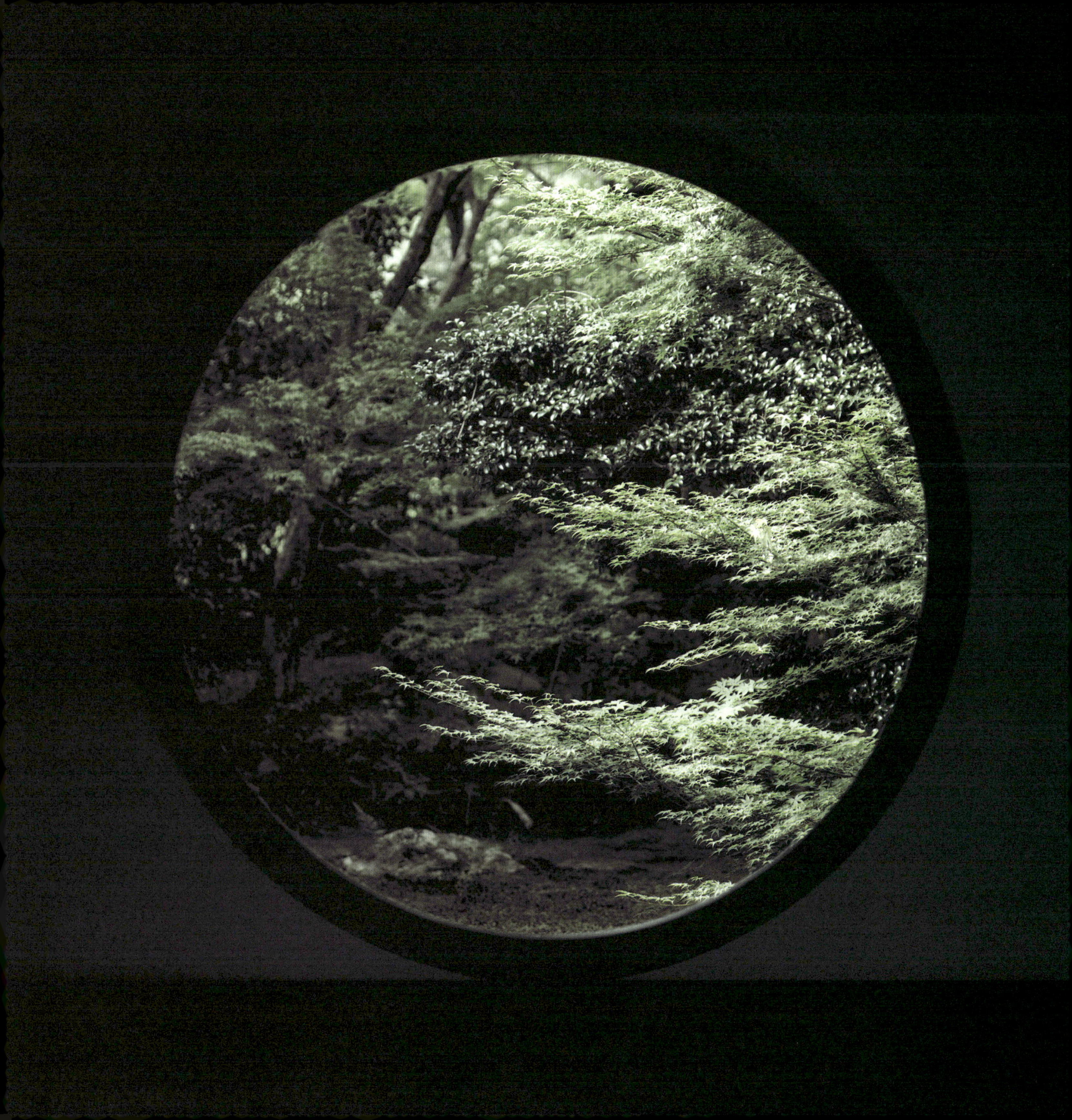

AN INVITATION FOR IMMERSED DELIBERATION

It is said that 2,500 years ago, *Shakyamuni*, the historical Buddha, was asked by his followers about the essence of his teachings, in reply he slowly twisted a flower between his fingers with a slight smile...and one of his students returned this smile!

Since that time, understanding through direct transmission from mind to mind, from heart to heart is considered to be the underlying pillar for perpetuating Zen Buddhist traditions and primarily also for the actual practical application of Zen teachings.

Based on the insight that essential things can only be understood beyond explanations, that words are inadequate, and terms and concepts are simply insufficient when it comes to what is significant, Buddhism in general, but the Zen schools in particular, focus greatly on meditation as the most important tool for gaining awareness.

In the daily monastery life, the meditative approach is applied to all activities, even the most seemingly insignificant of them, which are all regarded as opportunities to practice focused attention.

The premise for this is that, in principle, all aspects, manifestations, and constellations of existence contain comprehensive opportunities for gaining awareness, at all times!

To face these, in a confronting manner, so to speak, the Zen code of practice often applies paradoxes to take the seekers beyond their concept-based, conditioned thinking to actual experiences and direct comprehension, to realize the real nature of their existence through fundamental insights.

When they truly deserve the name, Zen gardens are created and maintained based on this mental attitude, and can actually only really be understood based on it. Therefore, the specific characteristic of Zen gardens is the fact that they are sites that are explicitly born out of spiritual intuition, are maintained and developed further in this spirit (which highlights the process-based and open nature of this approach as well as spiritual training on attentiveness), and finally to allow visitors to not only enjoy the, always temporary, result, since a garden is never really finished, but rather to actually participate in this creative process.

This involves more than the mere planning and creation of these gardens – as the frequently drastic changes that they go and grow through repeatedly require concentrated awareness, inward looking, as well as the greatest possible openness among those who diligently care for them and carefully develop them, giving them a future.

Often, experienced monks were entrusted with the creation and design of the gardens and the maintenance itself was considered a spiritual practice within the monastery context. Today, this is still partially the case, even though increasingly professional gardeners are handling the care of the gardens, as here also there is a lack of young monks and nuns to carry on the legacy.

Times are changing after all, but they always have: Looking closer at the history of individual gardens, it is again and again surprising to see the often turbulent destinies of some gardens across centuries that transformed them again and again.

For example, some temples were originally based on villa grounds that were created by, often artistically inclined, aristocrats as retreats or that served as remote summer residences for members of the extensive imperial family, or as retirement residences after abdication. Later, these were often dedicated into temples.

These, in turn were often involved in various strives and power struggles, etc., which frequently led to changes of Buddhist schools of thought as well as more or less drastic losses of the grounds held by the temples, up to and including their total destruction. In particular the often devastating fires during the long periods of war in Kyoto inevitably led to the creation of new garden sites along with in part extended periods of neglect. Today's visitors of, in part world-famous, gardens cannot even guess are the deplorable state of quite a number of these gardens only a few decades ago, considering how perfect they look today, seemingly unchanged for centuries. However, this is deceiving, making it even more important that some courageous persons today who are aware of both the depth of traditions as well as the conditions of modern times assume the care of these gardens based on an adequate spiritual approach!

Similarly, this also applies to today's visitors – the true essence of these gardens is discovered in line with one's own state of mind, the unbiased and ideally unlimited openness, and the readiness to actually become involved with these fascinating universes, to open up to them without any reservation, to delve into their serenity and meditative lingering.

In Japan, creating gardens within and based on a spiritual context has a long tradition going back to practically mythological times. Consequently, ancient, in the broadest sense garden-like sites that are often nothing more than enclosed open spaces or simple markers indicate the desire of their creators of establishing interactive sites. These are often located near century-old giant trees, exceptional rock formations, impressive waterfalls, the estuary of rivers, or other remarkable places that are frequently quite remote and located, even in this day and age, in the middle of wild nature, which of course was much wilder a couple of centuries or even millennia ago!

The *kami* (often rather ineptly translated as "gods", but rather spirits) are believed to reside in these special locations, while the open spaces are an invitation to them to come by every once in a while, even if only temporarily.

This was probably also fueled by the wish of locating these spirits, to venerate them, to celebrate rituals for them, possible also ask them for assistance, but at any rate to pay them respect.

The world vision of the ancient Japanese was distinguished by great respect for these *kami*, which incidentally are always thought of in the plural as they number several thousand!

They are the ones who determine the fate of the world and thus the destiny of the people in it.

The deeply-felt respect of the elements of nature and the forces that shape them was probably so deeply engrained in the minds of the residents of this island nation based on their feeling of being particularly vulnerable to the, frequently brutal, forces of nature.

This is because in Japan, apart from typhoons as well as extreme heat and cold waves, volcanic eruptions, earthquakes or seaquakes, followed by occasionally very destructive tsunamis and fires are much more of an everyday reality than elsewhere.

The result is an increased collective sense of fragility, of awareness of the ephemeral nature of all constellations and phenomena, and thus a peculiar sensitivity in dealing with nature, its manifestations, and the above-mentioned *kami*, the spirits of nature...

One outcome is an almost unconceivable variety of, often much localized, customs, rituals and celebrations all across Japan. Many of these have long since been forgotten, some however, only increasingly diminished in recent times with the remaining ones threatened to disappear altogether or to continue to meander along for a short while as mere imitations devoid of any deeper meaning...

Nevertheless, some of these traditions continue to survive today as a vaguely and loosely assembled conglomerate that is known as Shintoism, generally referred to as the original religion of Japan. Yet this categorization is more than questionable and can be seen only as an attempt to distinguish these traditions from all influences that were introduced to Japan from the outside world in the course of history, and primarily to create a national identity momentum in the late 19th/ early 20th century. This instrumentalization and simplification does not do justice to the many centuries of interconnection and mutual interactions of a great variety of spiritual directions and customs in the country.

Yet what is essential is the fact that this attitude of almost instinctive respect of nature was the determining force for all cultural historical developments, including the various directions of the gradually evolving landscape arts in Japan.

Another key characteristic of some Japanese, especially among those dedicated to the arts, which clearly originated from the respectful interaction with nature, is the pronounced tendency to dedicate themselves to highly concentrated and thus in-depth contemplation

of natural phenomena, to virtually immerse themselves in them, resulting in extreme cases in a transcendental unity with them.

Vivid examples of this long, close and very deep relationship between the Japanese people and nature are the top artistic outputs that still touch us in an odd manner today, such as the traditional ink drawings of natural subjects or the art of poetry.

In some of those works a degree of understanding appears to have been reached that is less based on knowledge and skills, instead going far beyond that into the realms of intuition and meditative experiences of merging and "becoming one". These instances result in works that are surprisingly capable of transcending time, space, and cultures to communicate essential key information in an abstract form.

Here also, therefore, it is the state of mind that is essential and at the root of practically all traditional Japanese art forms and especially those inspired by Zen Buddhist thought and the associated practical application of its principles. This may perhaps be experienced most directly in Zen gardens, of which Kierok presents some of the most outstanding examples in this book by means of his subtle visual language.

To be able to understand the cultural historical background of these gardens it may be helpful to know that starting in the 6th, and increasingly from the 7th century and during the following centuries, Japan voluntarily sought, supported and allowed the export of Chinese culture, mostly via Korea, in several waves. The dimensions and implications of this globally unprecedented approach cannot be overestimated.

These massive influences now met with the so to say indigenous Japanese traditions that had previously developed, which they absorbed, transformed and assimilated in waves. As a result, continued periods of reorientation sustainably affected all areas of life in Japan, if not forming the basis of almost all aspects of what we today consider traditional Japanese culture.

In addition to Confucian and Taoist thought, this enormous cultural import included Buddhism, which increasingly gained in importance after the courtly and clerical social structures had developed in Japan, also based on the Chinese model. A more profound effect on the arts, including landscaping, only emerged later.

The esoteric schools of thought in particular that were developed relatively late within the Mahayana Buddhist context and that came to Japan from the 9th century onwards (especially Shingon and Tendai) initially led to the development of temple gardens representing the so-called "western paradises" of the Amida Buddha. The height of this very symbolic and rather lush garden style was reached in the 12th century and largely dominated the gardens of Japan not only within the Buddhist religious context but also to a large degree the aristocratic and courtly world.

Starting in the 13th century, the Zen schools arrived in Japan from China. These not only caused a profound restructuring of Buddhist thought and spiritual practice, but had a direct and indirect impact on almost all aspects of Japanese art and culture that cannot be overestimated.

The stringent approach of Zen, which was highly valued and therefore promoted by the aristocratic warrior caste, led to the development of an entirely new esthetic approach that largely became the basis of Japanese art and culture and that seems to continue to resonate to this day.

The gardens that emerged from the Zen spirit are distinguished by their iconic sparse rock placements on raked gravel surfaces, only intercepted by limited perennial bushes or mosses. These gardens often contrasted with the powerful natural wilderness that used to surround them. For the most part, we can only guess at this today because, especially in Kyoto, most temple sites, as remote as they may have been at one time, have long since been incorporated into the busy city, while the majestic mountain ranges that used to extend into the distance are today partially built over, interrupted by power lines, and planted for long stretches with uniform cedar forests...

Nevertheless, what remains is a carefully composed multifarious network of counterpoints, which amazingly enough manages somehow to convey tremendous tension while emitting great and incredibly deep serenity.

These meticulously balanced and symmetrical energies are possibly the source of the incredible dynamic power that is inherent in the gardens and that they radiate.

Even though the compositions of Zen gardens, especially the rock placements, also contain symbolic and mythological references (which were admittedly often only apocryphally attributed to them), their creators rarely intended to stage idealized imaginary worlds in reference to real or mythological landscapes of Japanese or classical Chinese ink paintings, or to evoke paradise worlds such as the transcendent Amida Paradise Gardens mentioned above.

Rather, Zen gardens are conceived as meditation areas and as an invitation to those meditating in them to directly experience the universal teaching of Zen Buddhism. For example, a connection can be found between the carefully composed and regularly harked gravel areas that often play a central role in Zen gardens and whose abstract shape exudes a certain degree of barrenness yet paradoxically seems to vibrate with vitality, and one of the central Buddhist teachings: to detect emptiness in all phenomena, which simultaneously indicates the limitless potential of all existence.

Similarly, with the approach of focused immersion in view of such a carefully composed garden, aspects of the ephemeral and constant change become obvious. The meditating individuals find themselves directly confronted with constant decay and new growth in and all around them. The inseparable connections among all these multi-layered interwoven processes and the forces underlying them can be immediately experienced by the attentive mind.

Although the basic elements of the "dry landscape gardens" were present long before, they only became an independent garden style around the *Muromachi-jidai* (from around the 15th century), gaining increasingly in importance then coming into full bloom to eventually become one of the most dominant traditional independent garden styles.

The traditional name for these extremely reduced dry gardens is *karesansui*; the Japanese characters of this word stand for dry, mountain/s and water, whereby the rocks stand for mountains, thus representing one of the cosmic elemental forces – *yang*. On the other hand, the dry, usually white colored gravel that is harked in wave-like structures symbolizes water and represents the dynamic *yin* principle. In this sense, universal principles are reflected in the garden microcosm.

Similar to cosmic dimensions, in the gardens the basic premise is to find subtle balances on multiple, mutually dependent levels, while at the same time everything is perpetually in motion and changing. Or rather it can be said that this is precisely their underlying dynamic.

These dry gardens are usually found surrounded by other types of gardens, such as those inspired by classical Chinese iconography, or others that reference real or idealized landscapes, frequently featuring ponds, watercourses or waterfalls. This attractively highlights the high degree of abstraction of the *karesansui* even more.

The surrounding gardens are also often intended for strolling through, which is another difference from meditation gardens that are mostly composed as pure contemplation gardens.

While strolling gardens are discovered in movement, *karesansui* are focused on a very specific fixed point, which is located in the veranda, but even more frequently somewhere inside the adjacent temple building.

If one assumes a seating position (which is the eye level towards which the entire site is composed), a landscape image emerges that is framed vertically by the pillars carrying the canopy, the overhanging straight edge of the roof on top and the horizontal structure of the veranda on the bottom. This framed excerpt not only defines what the observers are allowed to see, but (even more importantly!) precisely what cannot be seen from this position.

Another interior/exterior interaction results from the relation of the *karesansui*, which is often framed by a wall that may enclose the actual dry garden, to the world beyond this boundary.

This world may consist of surrounding mountain ranges and lush forests, or of other gardens, frequently belonging to the temple grounds, for example consisting of tall bamboo groves or of striking, sublimely aged tree giants, or even a gracefully swaying and drooping Prunus Mume tree, whose fragile or wild vitality forms energy counterpoints.

These are referred to as *shakkei*: borrowed landscapes, which although only indirectly belonging to the actual garden or sometimes even located far away, are nevertheless very essential for the balance of the entire composition and its overall effect.

From an esthetic point of view, the extremely complex principles – also relevant in other traditional arts, especially those based on Zen, such as the dynamic balancing of forces by means of asymmetries while keeping them oscillating, play a central role. These principles also include the emphasis on open spaces (corresponding to the carefully placed empty areas in classical ink painting, or calligraphy ...), or the inclusion of, presumed, imperfections.

All these aspects, which are sometimes harshly contrasting and sometimes subtly interplaying, should be coordinated with each other applying the most unpretentious understatement and simplicity possible, and carefully balanced.

Taking a closer look at these aspects and any attempt to implement them, will quickly reveal the above-mentioned contradictory nature of these requirements – that is their basic principle!

For example one requirement is that a Zen garden should appear totally natural, while in practical implementation almost nothing is left to take its natural course...

Yet the most outstanding compositions manage to fulfill even these requirements and are able to reconcile all these seemingly contradictory aspects with an almost floating and virtually vibrating balance to naturally be at peace within themselves and to exude this inner peace.

The gardens then invite visitors to discover this peace within themselves. This is because they are not intended as a counterpart to be perceived by the recipient as a three-dimensional picture, to enjoy and maybe attempt to decipher, but rather as a concentrated field of energy that actually allows visitors to experience all-encompassing reality as such.

Even though it may be helpful and interesting, possibly even illuminating, to approach the magical worlds of Zen gardens by applying cultural-historical classifications or creative categorizations, this can only illuminate the contexts and connections to a certain extent. When it comes to the essence, however, they not only fall short, but are actually contrary to Zen Buddhist thinking to a certain extent.

For Zen is precisely about of freeing oneself from categorizations; of breaking through conditioning; of clearing the field, of becoming open to unbiased experience in itself!

Therefore, it seems to me that the decisive, century-old relevance of Zen gardens lies, after all, in their spiritual dimension, and their very gentle invitation - just like Kierok's visual worlds – to pause and experience realities.

It remains up to us to follow these invitations.

Bernd A. Schellhorn
Translated by Cosima Talhouni

CONTRIBUTORS

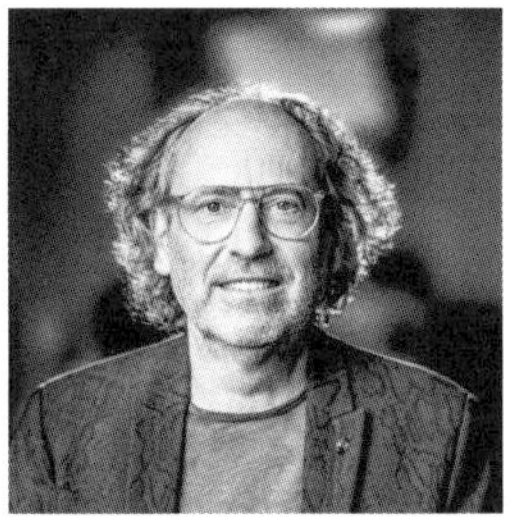

Thomas Kierok

A fascination with people led Thomas Kierok, after his training at the Lette Verein in Berlin into portrait photography. In his images, he succeeds in capturing what is special about each person. As a freelance photographer for over 25 years, he has been portraying ordinary people as expressively as celebrities from Karl Lagerfeld to Angela Merkel. He is represented by the renowned agency laif. His clients include magazines, advertising agencies, and companies. Kierok also works as a lecturer and increasingly devotes himself to artistic projects.

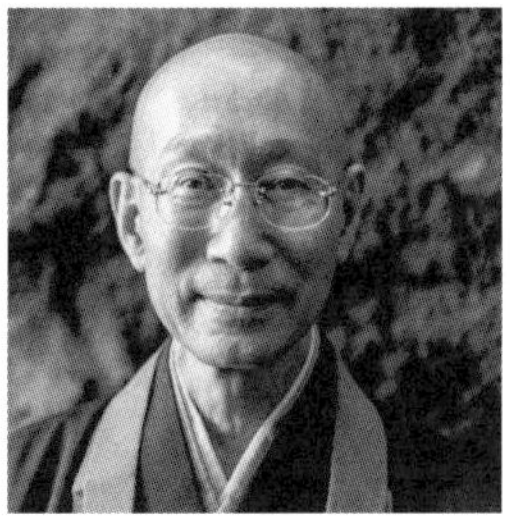

Masuno Shunmyō

Born February 28, 1953 is a Japanese monk and garden designer. He is chief priest of the Sōtō Zen temple Kenkō-ji, professor at Tama Art University, and president of a design firm that has completed numerous projects in Japan and overseas. He has been called "Japan's leading garden designer."

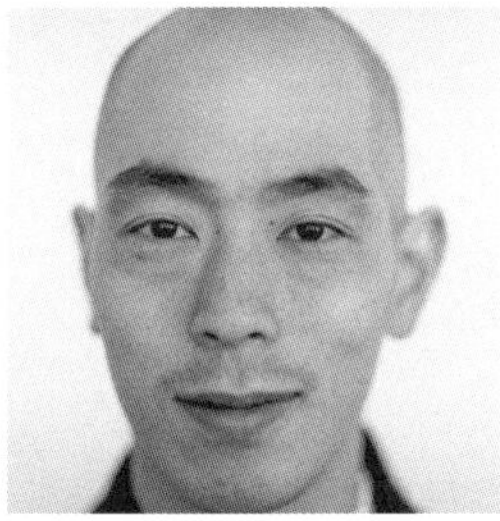

Higuchi Seigaku

is a monk. After graduating from university, Seigaku practiced at Eihei-Ji, the main Zen temple in Japan. During his practice and afterward, he traveled not only throughout Japan but also to temples around the world to show how to apply Zen to modern life.

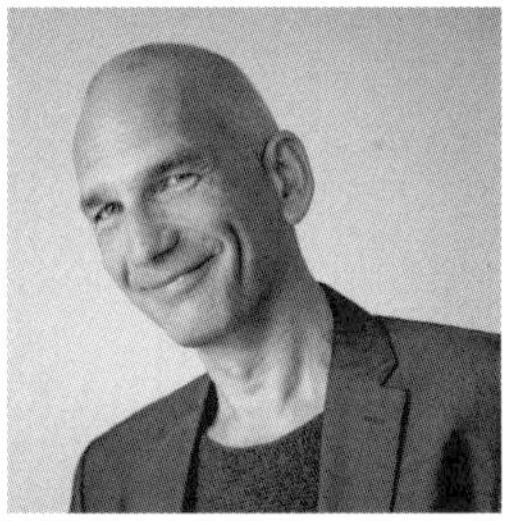

Bernd A. Schellhorn

A curious interest in traditional Japanese art and Zen Buddhism initially led him to Japan in 1990, where he studied traditional ceramics and related arts, such as tea & ikebana for five years under accomplished masters. Back in his native Germany – after studying art history and art management – he worked for some of the most prestigious international contemporary art galleries in Berlin and New York for about a decade since 2007. Based in Japan and Germany, he is an independent art consultant focusing on contemporary international art on the one hand and Japanese art and antiques on the other, occasionally designing and leading exclusive, customized art study tours to Japan for Western art connoisseurs and collectors.

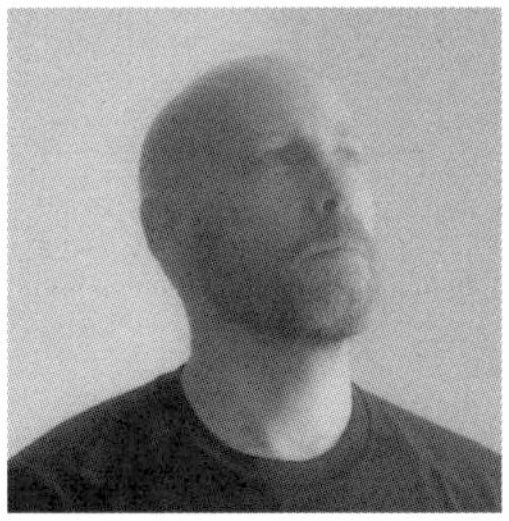

Benjamin Wolbergs

is an art director and editor who lives and works in Berlin. His work as an art director includes over 100 books for international art book publishers. As an editor he is eager to find new inspiring themes in the field of art and design that have not yet been published. His highly acclaimed anthology *New Queer Photography* was published in 2020 by Verlag Kettler and Gingko Press and has been awarded with Silver by the German Photo Book Award 21/22 in the category "conceptual-artistic photo book". (www.benjaminwolbergs.de)

LOCATION INDEX

The Deutsche Nationalbibliothek lists this publication in the Deutsche Nationalbibliografie; detailed bibliographic data are available on the Internet at http://dnb.dnb.de

ISBN 978-3-7165-1872-4

www.benteli.ch

1st edition 2022

Artistic support: Oyama Yuka
Texts: Masuno Shunmyō, Bernd Schellhorn, Nagase Mitsue, Thomas Kierok
Translation (Preface Masuno): Nagase Mitsue
Introduction (Interview und Text): Ursula Richard
Calligraphy: Higuchi Seigaku
Art direction: Benjamin Wolbergs
Editing: Cosima Talhouni & Friederike Christoph
Author's photo (Bernd Schellhorn): Nagase Mitsue